MAPPING
OUR
GENES

LOIS WINGERSON

MAPPING
OUR
GENES

THE GENOME PROJECT
AND THE FUTURE
OF MEDICINE

DUTTON  NEW YORK

DUTTON

Published by the Penguin Group
Penguin Books USA Inc., 375 Hudson Street,
New York, New York 10014, U.S.A.
Penguin Books Ltd, 27 Wrights Lane,
London W8 5TZ, England
Penguin Books Australia Ltd, Ringwood,
Victoria, Australia
Penguin Books Canada Ltd, 2801 John Street,
Markham, Ontario, Canada L3R 1B4
Penguin Books (N.Z.) Ltd, 182–190 Wairau Road,
Auckland 10, New Zealand

Penguin Books Ltd, Registered Offices:
Harmondsworth, Middlesex, England

First published by Dutton, an imprint of Penguin Books USA Inc.

First Printing, August, 1990
1 3 5 7 9 10 8 6 4 2

Wingerson, Lois.
Mapping our genes: the Genome Project and the future of medicine /
Lois Wingerson.—1st ed.
p.   cm.
Includes bibliographical references.
ISBN 0-525-24877-3
1. Genetic disorders—Forecasting.   2. Chromosome mapping.
3. Human genetics.   I. Title.
RB155.6.W56   1990   89-48560
616'.042'0112—dc20   CIP

Printed in the United States of America

Set in Palacio

Designed by Steven N. Stathakis

For my mother,
Leslie Eitzen

CONTENTS

ACKNOWLEDGMENTS

A project of this magnitude on this topic is a humbling prospect, especially for a nonscientist. I owe my thanks to all those who did not discourage me from taking it on in the first place, and since no one discouraged me, I should therefore thank everyone who heard about the concept. I am especially grateful to Phillip Sharp of the Massachusetts Institute of Technology, whose early encouragement helped me to push ahead.

It will be obvious to scientists and others familiar with genome research that this is not a comprehensive account. Nor was it intended to be. *Mapping Our Genes* was written for the many members of the general public who, by 1990, were still quite unclear about what the genome is, much less the implications of mapping it, at a time when policymakers and scientists had long since set the process in motion. My highly selective approach was designed from the outset to entertain people as it informed them about these matters, especially as they relate to the crucial issue of finding solutions to human disease. I focused on mapping rather than on sequencing because what was being generated in the late 1980s was, most visibly, a map.

By nature, I used the approach of a journalist, not an academic. I traveled to the places where people live and work, and interviewed them. I watched research in progress. I attended numerous scientific conferences and studied the most relevant reports in the scientific literature relating to the topics I had chosen.

Several people did advise me to narrow my focus to a single research group involved in gene mapping, but I felt strongly that this would be unwise and unfair. The very essence of research into our genome is its great diversity, and I wanted to communicate that by portraying several different research teams and a variety of approaches.

My choices of topic were arbitrary and somewhat quixotic, and there are certainly hundreds of people I could have chosen instead. My sincere gratitude to the scientists and doctors portrayed here, extremely busy people all of them, who spared me time and attention, and patiently answered ignorant and perhaps sometimes insulting questions. I am especially grateful to Cassandra Smith for granting me complete freedom in her laboratory at a difficult time, and also for her candor. I owe thanks and apologies to several people who gave me considerable time and who ultimately did not appear in these pages, notably Drs. Lyle Best and Demitri Papolos.

The public lives of scientists are one matter; the stress and heartache of private individuals faced with disease are quite another, and I am as much astonished at as I am grateful to those families who let me intrude into and write about their lives. Those who are named are named; those who are not know who they are. I am especially indebted to a few people who will probably never read these words: those numerous members of the Old Order Amish community in Lancaster County, Pennsylvania, who showed me the meaning and nature of *Gelassenheit* and spoke to me candidly and willingly once I promised them anonymity. I shall always have warm and grateful memories of the Amish grandparents who let me live within their household for a few days.

Above all, I'm indebted to the clear vision and blockbusting genius of my remarkable editor, Meg Blackstone. I hope this book will have many more readers than Meg, but it could hardly have a better one. It would never have come into being without the wisdom and encouragement of my agent, Faith Hornby Hamlin.

Several other individuals suffered through early drafts of this book and commented on them. My sincere appreciation to Mark Fuerst, Janet Offensend, Pamela Pearce, Steve Ringquist, Elizabeth Ross, Abigail Wender, and particularly my husband, Mark. Laura Eitzen made the sketch illustrations.

Finally, my deepest gratitude to my family—Mark, Alexander, and Kate, and my mother, Leslie—for their long-suffering encouragement and faith in me.

Betty LeBlanc's ancestors, her father's family at top and her mother's family above. (The woman with the blanked-out face refused permission to be identified here because she felt it might ruin her children's chances of marriage.) Courtesy of Betty LeBlanc.

1

BETTY'S KINDRED

Quite suddenly, in the last decade, biologists began to spell out human nature in an entirely new way. At a breathtaking pace, they have proceeded to characterize it in nature's own terms, in the language of molecules we call genes. Their research began with the study of heredity, by observing people within their families.

Every family has its stories to tell, and others that it keeps secret. Another kind of story is written in our genes. We can read parts of it by looking at each other, but most of the details are obscure.

Usually, the specifics of any family's genetic heritage are blessedly mundane. Sometimes, however, they reveal a story so momentous that it redefines a person's concept of who he is.

For a few weeks, Darren LeBlanc went jogging. He would rise before dawn and go out behind the barn, rousing the chickens and cows, to the grazing pasture. In the near darkness he would begin to lope around the perimeter, through the deep shadows of loblolly pines as tall as ten men.

After a short time, always, Darren would stumble. After another fraction of a mile, his legs would give way and he would fall into the damp grass, and then get up and go on. He would continue to trip over nothing but shadows. But he would continue to jog.

With enough running, Darren had decided, he might grow strong and gain weight. He would not be so puny and ugly. You can't run fast, he thought, because you're not in shape. You can't do anything well. Get in shape. Run. Again he would fall.

Darren's mother might argue with the term "puny," but Darren was right about his size. He was a delicate boy, small and slender. At fourteen, lean and wavy-haired, he looked more like a poet than like the farmer he expected to become. But he was not ugly. He did not look unpleasant, or even unhealthy.

It was also untrue that there was nothing he did well. Anything he planted would grow. He was in charge of the cucumbers in the garden, and they flourished. His other duty was to care for the chickens. He fed them reliably and kept the coop clean.

He followed his grandmother around when they visited her, eager to know everything she knew: How to set the incubator. How to make soap. How to clean a chicken. Darren also did very well in school. He read eagerly, especially about animals.

His savvy and considerable courage made up for his fragile appearance. If someone beat him up, he might tell the bully he'd keep quiet about it, in exchange for some possession. But kids had subtle ways of getting back at him. All they had to do was take Darren along somewhere and then run away, leaving him to wander back alone. He'd never catch up.

Quite early in his life, Darren decided something must be wrong with him. It seemed he could never carry a bucket of eggs without dropping it, or a pail of water without spilling. His brothers could heft a fifty-pound bag of feed and pour it out, but Darren couldn't lift one on a dare. Darren compensated for his clumsiness by being careful and slow. He carried buckets of water slowly. He washed dishes slowly. He ran slowly.

During junior high, Darren briefly considered an alternative explanation for his problems: he might just be out of shape. Jogging was the experiment that disproved the alternative theory.

"I could never improve," he recalled later. (The syllables halted briefly, one by one, on his tongue.) "I'd get in shape that much, you know, I could do the same things every day. But I would never improve until, finally, it was almost impossible to run. I would fall so much, and I'd end up just hurting myself. I'd have so many scratches and bruises that it was just useless to run anymore." Years after the fact, he told the story with a little smile, as if his foolishness amused him now.

Just before he entered high school, Darren took his first trip away from home alone. He joined a youth tour of East Coast cities, sponsored by the United Nations. In New York, he visited the Statue of Liberty and, of course, the UN itself. There, for the first time, he saw images of people starving. He was most affected by a photograph of a little girl selling cow dung to buy food.

The photographs jarred his view of the world. It struck him that people far away would die if they did not get help. Immediately Darren decided what to do with his life. He would become a doctor and do something to help other people, far away.

Meanwhile he was busy assuring people that he didn't need help himself. When the group climbed the Statue of Liberty, Darren nearly collapsed. Other teenagers on the tour noticed that he was walking slowly and sometimes bumping into things. They'd ask if something was wrong.

The last thing Darren ever wanted was for people to think he was different. "Oh, no." He would chuckle. "It's just because I wear boots too much." In fact, Darren never wore boots.

When his parents picked Darren up in Baton Rouge after the trip, they were startled to see him looking so gaunt and thin. He sank into the backseat, and slept all the way back to Orange, Texas. When they reached home, he staggered toward his room, fell onto the bed, and slept for most of the next three days.

"God, Derald!" Betty LeBlanc said to her husband. "He's so worn out! Don't you think something must be wrong with him?"

"Naw," said Derald, who is an easygoing and sensible man. "He's been gone for two weeks with a bunch of teenagers. He's just exhausted."

Betty might defer to Derald's opinion, but she did not share it on the subject of Darren. Her husband thought Darren was just lazy and a bit clumsy. But as Betty often said afterward, a mother can tell when there's something wrong with her child.

Betty's suspicions began when Darren was about seven, on account of his chores. Darren was aggravating. At some jobs he was scrupulous, even finicky. At others, he was a bungler. Betty asked the pediatrician about it.

"Don't worry," he assured her. "Darren is just smart enough to try to get out of his work. Don't pamper him."

"That may be true about work," she said. "But Darren gets tired when he plays! And what child ever says, I'm tired, and quits playin'? Never!"

When Darren was nine, Betty decided he might be anemic, and insisted that the doctor do a blood test. The results were normal, and again he told her not to worry. For five years, still uneasy about her middle son, Betty held her peace.

Two weeks after Darren returned from the UN trip, Betty found him in tears after swimming class, clutching his ears, saying that his head hurt. She helped him into the car and drove home.

"I think Darren has an ear infection," she told Derald. "I'm gonna take him to a doctor."

"To the pediatrician?" Derald asked.

"No way! I am not gonna fool with that doctor."

She got an appointment for three days later with an ear, nose, and throat specialist in nearby Port Arthur. With Darren and four-year-old Donna, the LeBlancs drove to his office. When they arrived, Betty took Donna to the bathroom and therefore missed the beginning of Darren's exam.

When she entered the examining room a few minutes later, Darren was on his feet, trying to walk with his eyes closed. He was swaying badly, supported by the doctor's hands at his waist. It was the most frightening sight she had ever seen. What about Darren's ears?

"Excuse me," the specialist said, as he helped Darren into a chair. "I need to make a phone call." As he passed Betty at the door he added, "Have you ever considered seeing a neurologist?"

"Why, no!" she replied. "No one ever told us to!"

The doctor called a neurologist in Beaumont, less than an hour away. "He will see you right away," said the doctor as he hung up, which frightened Betty even more. They drove immediately to Beaumont. As soon as they arrived, the neurologist asked Darren into an inner room. Betty and Derald followed.

The exam was brief. First the neurologist asked Darren to touch in turn his own nose, then the doctor's finger, as the finger jumped about in front of him. He scratched the bottom of Darren's feet, and held a tuning fork next to each ear. He tapped Darren's knees and elbows with a rubber hammer for reflexes, which were absent. And he asked Darren to stand, eyes closed and arms crossed. It's normal to sway and fail eventually at this exercise, but it utterly defeated Darren; he would lurch immediately, and open his eyes.

Afterward, the neurologist ushered the family into his office for a consultation. Betty remembers it vividly.

"He said, 'Well, I really think your son has Friedreich's ataxia.' " She repeats his words brightly, as if reciting a nursery rhyme. "Of course we wondered, What the heck is that? I said, 'Well, what is it?' He said,"—she makes him sound as cheery as a kindergarten teacher—" 'Well, it's a disease of the nervous system. The nerves deteriorate.' He said, 'Darren has lost a lot of the sensation in his feet already.' "

The neurologist went on to say that the disease is hereditary and very, very rare. Beyond that, he added, little is known about it. "To this day," he went on, "as far as I know, there is no cure."

"I said, 'Well, then, what's gonna happen?' " Betty recalls. "And he said, 'Well, it's gonna keep going up in his legs.' I said, 'Well—well, then, what's gonna happen?' And very, kind of, cold . . . he said, 'Well, then he just won't be able to walk anymore!' "

There was a very long silence.

"We definitely want a second opinion," Betty said at last. Then the neurologist looked directly at her and responded, for the first time, with great kindness. "I really understand," he said. "I understand. And I don't blame you."

Before they left the neurologist, Betty asked him how to spell Friedreich's ataxia. He had to take down a reference book to look up the answer.

During the hour's drive back home, the family was silent.
But after they got home, Darren's father approached him. "Well,
son," he asked. "What do you think about all this?"

"I don't believe it's true," Darren replied.

"Well, Darren, I kind of agree with you," Derald replied.
"But everyone else feels like this is very serious."

On the whole, however, Darren did not take the news seriously.
It was easy to pretend the doctor's visit had not happened. The
headache went away quickly, and his balance was no worse because of what the doctor said.

"I remember the doctor saying, when you're twenty you
won't be walking anymore," he says. "I kept thinking that,
Oh, as long as I stay healthy and eat right and just exercise
every day and stay strong, there's no way I'd get weak." He
bought some weights and began to do stretching exercises every
morning.

He also decided that at all costs no one else must learn about
the unfortunate incident. The doctors could be wrong (indeed,
they must be) or something might happen to change his circumstances. Besides, he did not want to give anyone another reason
to pick on him.

Shortly after his diagnosis, Betty heard about a self-
appointed "metabolic technician" who claimed to have saved
her own husband from cancer with high doses of vitamins and
coffee enemas. She said the treatment cleared the body of tox-
ins, and told Betty it could also help Darren overcome his disease. He was very eager to start the regimen, and Betty thought
it made some sense. At worst, she thought, it wouldn't hurt.
But Darren began to vomit every day and lose weight he couldn't
spare. After a few weeks, she refused to take him to the woman
anymore. Darren was furious.

He was also enraged that his mother was frustrating his
other plan, to conceal the diagnosis. She told anyone who would
listen about the illness. Friedreich's ataxia: the peculiar words
echoed in his mind. He grew sick of them.

People began to be very correct and careful when they spoke
to him. Conversations would stop in midsentence when he
walked by. And it seemed that every time he walked by, the
conversation was about Friedreich's ataxia.

A few weeks after the diagnosis, Darren heard his mother describing the condition to a neighbor. "If you notice, when Darren goes down the hall, he hits the wall like a drunk?" she said. "He sort of bounces from one wall to the other wall."

The description was ridiculous and, from a woman who would not allow liquor in the house, insulting. "I do not stagger like a drunk!" Darren shot back at her. "Why are you always picking on me? Look at Dana! She doesn't have good balance, either. I don't have good balance, but look at her!"

Betty gave a little start and replayed a mental image of her eldest daughter in motion. Darren was right: Dana's gait was awkward. She walked like a marionette controlled by a small child.

That evening Betty asked Dana to do a test she had seen the neurologist try on Darren. Standing cross-legged with her arms crossed and her eyes closed, twelve-year-old Dana could not keep her balance at all. She seemed not to have reflexes in her knees. Within a few weeks, the same neurologist who had seen Darren confirmed that Dana also had Friedreich's ataxia. It was too soon for Betty to discern that her second daughter, Dena, was also destined to develop the disease.

Betty had two immediate reactions to Darren's diagnosis: She grieved, and she couldn't believe it. When they stopped at a restaurant for lunch on the way home from the neurologist's office, she went straight to the rest room, locked the door, and sobbed. Then, as they drove home, she sat silent and thought about a young man at their church who had survived a near-fatal illness. We're gonna be like him, she told herself. We're gonna make it through this.

Until that point, Betty's view of the world had consisted of the beautiful, the wonderful, the real, real nice, and what she chose not to think about. What she had just learned about Darren did not fit into any of those categories.

When they reached home, Betty went straight to the encyclopedia. It took some time to find any reference to Friedreich's ataxia. What she found was brief, and misleading. The next sentence described ataxia as a late complication of syphilis.

Betty closed the book and immediately drove to the pediatrician's office, leaving Derald with the children. The pediatri-

cian was surprised to see her sitting, alone, in the waiting room.
"Well, Mama, what's wrong with you?" he said, patting her on
the thigh.

"Have you ever heard of Friedreich's ataxia?" she asked.
He went pale.

"Yeeeeees. Why?"

"Darren has just been diagnosed with Friedreich's ataxia."

The pediatrician sat down. "Wow," he said. "I don't believe it."

Darren's pediatrician had only seen one other patient with
Friedreich's ataxia in his life. Like the best experts on the disease, he knew little about it beyond the descriptive. It is in fact
quite rare: There are estimated to be roughly five thousand cases
in the United States (but frequent misdiagnosis makes any guess
likely to be inaccurate). At that level, hemophilia is four times
more common and sickle-cell anemia ten times as prevalent.

Friedreich's ataxia is named after one Nikolaus Friedreich of
Heidelberg, Germany, who first described it in 1863 (exactly a
century before Darren's parents were married). He gave about
as much detail then as anyone knew in 1981, when Darren was
diagnosed. *Ataxia*, from the Greek, means "without order or
arrangement" and describes the motions of someone with an
advanced case of the disorder. Friedreich's ataxia is progressive.
There is no cure. Even physical therapy is of little value. Its
victims die young.

The syndrome usually emerges sometime before puberty in
a previously healthy child, who gradually begins to stagger and
lurch while walking. Later his speech may slur and he will begin
to feel weakness in his limbs. Due to muscle atrophy, his eyes
may cross and his spine will probably begin to curve, as well as
his feet. The expected life span is shorter than normal, around
forty or fifty years (but not twenty or thirty, as Betty had been
told). Death usually results from heart disorders associated with
the syndrome.

Since 1863, the medical profession has honed its ability to
distinguish the symptoms of Herr Friedreich's ataxia from those
of a number of slowly disabling ataxias, such as Charcot-Marie-
Tooth syndrome, Werdnig-Hoffmann disease, and the most famous, Huntington's chorea. It has also learned from the study

of affected families that Friedreich's is hereditary, and also recessive. It must be inherited from both sides of the family. Each
parent carries one copy of the defective gene, but the gene is
"silent" in the parents, because each of them possesses a normal gene to counteract it.

Neither Betty nor Derald had Friedreich's ataxia; they must
each have one normal copy and one defective copy of the unknown gene. The affected children all inherited a defective gene
from each parent, and had no normal counterpart. Betty and
Derald were especially unlucky: The odds that any child will
inherit both defective genes are only one in four, yet half of their
six children developed the ataxia.

Medical science can deduce the heredity of Friedreich's
ataxia by observing who in a family is affected. But doctors do
not know exactly what is wrong in the muscles or nerves, or
how to help, beyond providing surgery for various secondary
problems and aids such as wheelchairs.

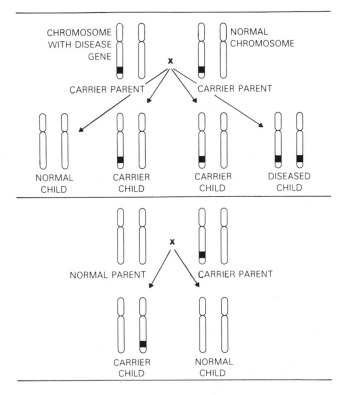

The recessive gene lottery

■   ■   ■   ■

It was two months before the LeBlancs were able to get a second opinion about Darren, but the delay didn't matter much. They expected the diagnosis to be confirmed, and it was.

The second neurologist tried to joke with Darren and treated him like a small child, making funny noises during the exam. Betty saw nothing to joke about. She had one important goal: learning about research toward a cure. There has been little, if any, the doctor told her.

"How can that be?" Betty asked, incredulous. "How can they not research something as terrible as this?"

"The disease is very, very rare," he replied. "You will have a difficult time finding another family with this.

"Besides," he went on, "how can we research the nerves?"

There was an awkward silence, and Derald rose to leave. "I'd like you to come back again in six months," the specialist said. "By the way," he asked Betty in passing, "are you and your husband related?"

"No-oh!" Betty said, taken aback. The suggestion was outrageous, preposterous. "Naaaw. Derald was raised in Orange and I was raised in Louisiana." The doctor didn't press the point.

The months after Darren's diagnosis blur in Betty's memory. Darren's problem engulfed her. She knows she didn't sing around the house or roughhouse with the boys as usual. She didn't care to eat. She didn't sleep much. Often, she'd get up in the night, wander into the living room, sit down in one of the recliners, and cry.

More than once, Derald followed her. "Don't do this, Betty," he would plead. "You can't do this. You're gonna get sick. Then what use will you be to anybody? You can't get sick."

A few months after Darren's diagnosis, Betty heard that a faith healer was coming to Lake Charles. She tried to get Derald and the children to go with her to the revival meeting, but only a friend of Betty's and her oldest son, David, would go. Darren would have nothing to do with it. When the collection baskets came around immediately after a few testimonials, Betty decided it was a fraud and left.

Then Betty proposed to fly to California with Darren, to see a doctor who claimed to have a drug that could treat ataxia. "How can they just give me a drug, if they don't know what's wrong with me?" Darren barked, and she dropped the idea.

Shortly after Christmas Betty learned about the National Ataxia Foundation and attended its next annual convention. From people at the foundation, she learned about a doctor in Denver who was doing biochemical research into Friedreich's ataxia.

The researcher did not encourage the LeBlancs to visit him, but they came anyway, twice, with all three affected children. It was the longest trip they had ever taken. He examined the children, confirmed the diagnosis, and told the LeBlancs about his research. He asked if Betty and Derald were related, and again Betty said no. He offered to see the LeBlanc children again if they wished to return, but he offered no new hope.

A highly qualified pediatric neurologist, the doctor had been on a fishing expedition of sorts. With no clue about the origins of Friedreich's ataxia, he had been examining skin-type cells from Friedreich's patients in search of any biochemical abnormality. He thought he had found one, and had published his results in a scientific journal. By the time of the LeBlancs' second visit, he was eager to look for abnormalities in a more relevant tissue than skin. He wanted to study the nerve or muscle tissue that must be affected in someone with Friedreich's.

He intended to obtain such tissue postmortem. As the LeBlancs were leaving, he asked Betty if she could help him locate someone with Friedreich's who had recently died. She had come to the researcher for hope; his last words to her mentioned corpses.

He was not being intentionally cruel. In fact, he could understand Betty's frustration. No one had been able to confirm his findings. To himself as much as to Betty, the research was a dead end.

In April, Betty took all the children except for twenty-two-year-old David (who was obviously unaffected) to see yet another specialist, who was opening a Friedreich's ataxia clinic at Tulane Medical Center in New Orleans. By that time she had her suspicions about her second daughter, Dena, and Dr. Michael Wilensky confirmed the presence of Friedreich's ataxia in all three children.

During the visit, Betty kept at Wilensky like a hungry bird: "Can't you do something for my kids? Why can't you help them?"

"Because I have no funds for research," he replied.

Wilensky is used to this kind of nagging. He told her what he told everyone grappling with one of the incurable neurological diseases: There's nothing I can do for the disease, but there's a lot I can do for you. You need some motivation. Help us get funds, or publicity. In the case of a rare disease, it takes extra effort to generate medical research.

"Why can't you start without the money?" she persisted.

"How can we start?" he replied. "We need a clue."

Before the LeBlancs left, Wilensky repeated the irritating question Betty had heard at nearly every consultation: Are you and Derald related? She responded as always, emphatically: No. That night, over the phone, she complained about this to her mother, Pearly.

"But we don't know, Betty," Pearly said. "We just don't know. After all, your daddy and I are related, seven generations back."

If there was nothing else to be done, Betty decided, she would do this: She would trace her line and Derald's back and out until she ran out of names. If the only thing that tantalized specialists about her family was the hint of a blood link between herself and Derald, she would either silence them for good—or else jolt them into action.

So it was that Betty LeBlanc became a genealogist.

Just before her wedding, Betty dutifully carried out a small obligation required by the Catholic Church. She verified that she was not closely related to her fiancé. For this purpose she paid a special visit to Derald's father, to ask the names of his forebears. Betty was pleased but not surprised to find that none of the names was familiar. They certainly did not appear in the Vincent family tree, which she had copied into her own large, white Bible.

Betty's hometown was Morse, Louisiana—little more than a rice mill and a train track across the road from a café and a gas station. Morse had a population of about five hundred. Roughly half of it was at her wedding. But her fiancé was a total stranger, a city boy from Orange, Texas, a hundred miles away, across a river and a state line. She had met him at a cousin's wedding. She admired his neat appearance and quiet, steady demeanor and also, later, his attractive offer of a way to leave Morse.

At the time, Betty was working as a beautician, still living with her parents in their small cottage that backs onto an irrigation ditch that drains a swamp. The Vincents are Cajuns, and their environment was typically Cajun, at the border between a swamp and a prairie, the two worlds that have been home for the Cajuns for two hundred years.

The Vincents' history was also classic Cajun. Betty's father, Nolan, like many Cajuns, had a life centered on rice and cattle. He kept cows (as well as some chickens and hogs) and worked in a rice mill. Pearly Vincent, Betty's mother, was at home in the swamp. Pearly's father was by profession a hunter, trapper, and fisherman. The family lived on alligators, ducks, garfish, and sugar from homegrown cane, cornmeal, and popcorn from homegrown corn.

Pearly's best dish was Derald's favorite and a Cajun classic, chicken gumbo, spiced with sassafras from a tree outside. "Whatcha got cookin'?" Derald would ask when they would visit her after the wedding. "There's a week I don't eat."

"Well, you can eat for a week here," Pearly would say.

At the time of her wedding, Betty paid little heed to her Cajun background. But it was pivotal to what would befall her family.

World history is told in books, but another account exists in the language used by the large molecule of heredity called DNA. The two histories always interact. The results of the interaction are seldom obvious enough to draw attention. In some families, however, these intertwined histories have had unmistakable consequences.

Put together, the genetic histories of many such families are beginning to reveal an even broader story about the entire human *genome*, the complete collection of human DNA. This larger story, once known only to molecules, is now being retold in terms our minds are capable of understanding. The implications for all of us are momentous. Scientists mean to learn everything written in our genes.

In Betty's case, the sad facts of Cajun history have one simple genetic tale to tell: the origins of Friedreich's ataxia. Betty LeBlanc traced that history in given names and surnames. Later, others began to unravel it in DNA.

The history in books records that the word *Cajun* is an Anglicized (and at first derogatory) permutation of the French term *Acadien*, which is what Betty's ancestors called themselves when they came to Louisiana in the late 1700s. The Acadians took this name in the mid-1600s, when they emigrated from France to what they called Acadia.

It was a region of the northeast coast of the New World including what the English thought of as New Scotland—Nova Scotia. Sharecroppers in both France and the British Isles had been offered land of their own if they would settle the distant and unpopulated region, but whose land it was to give was in dispute.

The French settlers cultivated the land and prospered in Acadia for more than a century. Acadia became populous—there were some ten thousand Acadians in Nova Scotia by 1755—but not by attracting new blood. At Port Royal, records show that nearly half of all marriages from 1729 to 1755 were between second cousins. In such a small, isolated place, everyone who spoke French was related.

In 1755, the British drove the Acadians into exile. The refugees scattered to the south, along the east coast of the New World. The Cajuns refer to the expulsion as "Le Grand Dérangement." Longfellow's 1847 poem "Evangeline" tells the story in English.

Betty LeBlanc's own ancestors, the Vincent kin, wandered unwelcome and homeless for thirty years, from slave quarters in Virginia to an English prison to a rocky island off the coast of France. Like the majority of the Acadian exiles, they eventually made their way to Louisiana, at a time when the Spanish governors of the region were looking for good Catholic farmers to occupy the land.

Many of the Cajuns settled in the bayous at the southern rim of the state and became fishermen. Others moved north into what would become the Cajun prairie, to farm and raise cattle. At the time it became a state, Louisiana adopted almost wholesale the parish divisions set up by the Catholic Church, and to this day refers to its counties as "parishes." Fittingly, the parish encompassing Morse, Betty's hometown, and most of the Cajun prairie, is called Acadia.

Most groups of immigrants to America rapidly set about trying to assimilate. Not so the Cajuns, by choice and by circumstance. First, they were physically isolated for several generations. Also, for reasons of language and by choice, they kept aloof from the people they called "Les Américains."

Today Cajun culture is still distinctive, in its food, its music, and its language. People in Betty's family speak two tongues: Southern-drawl English and the Cajun patois, which bears some resemblance to archaic French.

In Morse, the surname Vincent is pronounced "Vat song." Betty says her married name in the French way, "LeBlong," not "LeBlank." LeBlanc is one of the most common Cajun surnames; a character in "Evangeline" has the name, as does a large town in the region of France that Betty's ancestors left to come to Acadia.

The frequency of a surname in a population is "a better way to provide for remote consanguinity [inbreeding] than any other pedigree method," according to the eminent human geneticists Walter Bodmer and Luigi Cavalli-Sforza. Had she been a geneticist, Betty might have thought twice about the implications of Derald's surname, rather than taking at face value the quick search into the family Bibles before her wedding. As it was, she promptly forgot about it and went on her honeymoon.

Afterward, the couple moved into a brand-new bungalow in Orange. Derald returned to his job as a machinist at a chemical company, and Betty waited to become a mother.

Their first son, David, was born just over a year after the wedding. Darren was born the following year, and the year after that, Dwayne. In 1968, the family moved to a new house Derald was building on a two-acre plot just outside Orange, less than a mile off the freeway in a large grove of pines. Their first daughter, Dana, was crawling.

Before it was completed, the LeBlancs had acquired four acres of surrounding land—enough space to keep cows, chickens, and later guinea hens, turkeys, ducks, and horses. This was Betty's idea. The livestock would help to feed the family, she reasoned, and give the children useful chores to do.

If Derald had never learned farming, she figured, he must have it in his blood. Certainly Derald's father did; all his life he had saved to go back to the farm, and by the time Derald's house was completed, his parents had moved back to a farm in Louisiana, just outside Morse.

The land had been in Derald's family for generations, and was adjacent to his grandfather's farm. Derald's father had moved to Orange during the Depression, to work in a rice mill. Orange, Texas, was not his home at all. Morse was home.

There was good reason for this, according to Betty's cousin Clyde Vincent. Clyde is one of those "remembrancers" who exist in families all over the world, the person with a memory for names and an interest in history who collects genealogy and other facts about his ancestors, reaching back to time out of mind.

Whenever a group of Cajuns left Louisiana for Texas, he says, they did so because of some calamity. They moved because the rice and sugar plantations closed off open grazing lands for their cattle, or because boll weevil infestations killed the cotton crop three years running, or to find jobs in the refineries or rice mills during the Depression. Very few left their hometowns voluntarily.

"My parents and most of them [who moved to Texas] never really unpacked their bags," he says. "Their heart was in Louisiana. Louisiana was a place they did not want to leave."

Of course there are exceptions, people of Cajun blood who have left home for good. Over the generations, Cajuns have become doctors at big-city universities, war heroes, politicians. One Cajun, Ron Guidry, became a pitcher for the New York Yankees. Betty LeBlanc intended to put the Cajun prairie behind her, too. But in general, Cajuns born on the Louisiana prairie stayed close to home.

At the time of her wedding, Betty Vincent saw no need to trace the branches of both family trees any farther back or out than required to obtain the answer she was looking for: that she was not related to Derald. But the stranger she married was in fact the son of a neighbor in a small town that changed little from generation to generation. When Betty saw good reason to follow the family tree farther, she would discover a thicket; no, a bramble, with branches that intertwine, tangle, and, in places, rejoin.

As a nation of immigrants, Americans are among the first to be obsessed with what you might call "bourgeois genealogy"—tracing the roots of families not in the nobility or royalty. To the uninvolved, this obsession is difficult to understand. People who do their own genealogies find the work fascinating, but tedious. Betty's motives were different from most, but her methods were the same. The result, as often happens in genealogy, caused a mixture of satisfaction and dismay.

Betty began by asking Pearly and her mother-in-law everything they could remember about the family, not just about their parents but about cousins and aunts and uncles. With a list of family names and rough dates scrawled on odd bits of paper, Betty began to drive in the evenings to the public library in Crowley, which is the seat of Acadia parish, to search vital records.

The fact that Betty and Derald were Cajuns made the task much simpler. Going back all the way to Acadia, parish priests had kept detailed vital records at many stages of earthly life: birth, baptism, first communion, marriage, and death. In the 1970s another Cajun priest, Father Donald J. Hébert, undertook the massive effort to consolidate it all. He and an army of helpers compiled a comprehensive list of Cajun names and life statistics, drawn from church and courthouse documents in Louisiana and stretching back to Acadia. Once Betty discovered Father Hébert's volumes, she could trace almost any name that came up.

For several weeks Betty made the hour-long drive to Crowley and back in the evenings, after Derald had come home from work, to consult Father Hébert's books. After a few weeks, Betty learned that the volumes were also among the holdings of the Orange public library, nearer home.

By her second visit to consult them at the Orange library, Betty was beginning to feel fairly proficient. A clerk had shown her how to track marriage and birthdates and how to cross-reference them on special record sheets.

It was a Thursday, Derald's day off that week. She had been

in the library since morning, without a break and without lunch, following a particular trail of names that wound from volume to volume. A half hour before closing time, she suddenly leaned closer to the book. She was somewhere nine generations back among her husband's forebears. She had found a familiar name, where she did not expect it.

Quickly she leafed through her papers until she found the page that began with her own father. There it was: Ursin Dubois, and his wife, Celine Ystre. Betty turned back to the volume she had been looking at. Dubois, Ursin. Married to Celine.

She closed her eyes and leaned back in her chair, holding the page open. You're getting tired, she said to herself. Your mind is playing tricks.

A voice on the loudspeaker drew her back: the library was closing. Quickly Betty retraced the path through the volumes, the path that had led to Ursin and Celine. It led there again.

Betty drove home quickly, and plopped herself down in a kitchen chair. "Derald!" she said, out of breath. "Derald, you are not gonna believe it! We really are related!"

Derald looked at her wearily. The last of the six children had just gone to bed. "Yeah?" he said, without much enthusiasm.

"Well, wouldn't that interest a doctor a heck of a lot more than if we just went in and said, 'Hey my children have Friedreich's ataxia?' " she said. "If we went in and said, 'Hey, my children have Friedreich's ataxia and Derald and I are related.' Wouldn't they start to say, Wowwww?"

"They might could," he said quietly.

On impulse, Betty lifted the phone and dialed Michael Wilensky at home. It was past 10:00 P.M.

Wilensky was startled to hear Betty's voice so late in the evening, and sounding so chirpy. "That's great," he told her. "That's really something to go on! Keep going."

Ursin and Celine proved to be only the first of several ancestors she and Derald had in common. By a circuitous route of relationships, they were third cousins. Derald was also her sixth cousin in four different ways and her fifth cousin by two separate paths of descent. Betty and Derald were distant enough to satisfy the rules of the church, but not distant enough to escape accidents—accidents of history, accidents of nature.

Of course the next person Betty informed about her discovery was Pearly Vincent. Pearly's reaction was typically low-key, but surprising.

"Let me tell you something," she told her daughter. "All the Cajuns, they all related." As she remembers it, Betty laughed.

Pearly Vincent may be an uneducated country woman, but she has a deep faith buttressing a large measure of common sense. During the most difficult times, these qualities made her Betty's strongest human support. When Betty had a question almost impossible to face, Pearly would always come back, in her low, soft voice, with an answer.

Betty called her mother the morning after Darren was diagnosed, and told her the unbearable news. Betty said the doctors had told her Darren would lose the ability to walk, and would die young.

"Huh-uh," Pearly said. "Don't keep that in your head. The doctors don't know all. We just have to pray, and just take care of him. Do whatever we can."

When Betty discovered that Dena was also affected, of course the first person she told was Pearly. "This thing is supposed to be so rare, and three of my children have it," she said over the phone. "Three out of six. Why, Maman? Why?"

The question begged any satisfying answer, but Pearly responded nonetheless.

"God has chosen you and Derald to have children like that," she replied quietly. "God is testing you to see how well you will take care of them. What we all have to do is take care of them . . . and do all we can."

Because her memory often failed her since her stroke, Pearly wrote down the words *Friedreich's ataxia* the first time Betty spoke them to her. The next time she was quilting with her niece Vivian, she passed on the sad news about Darren.

The news tripped off a memory in Betty's cousin. "You remember those two boys who went to school with your kids?" she asked. "They used to walk so funny, and they was in wheelchairs when they left school?"

Instantly Pearly knew she had to find out whether those boys had the same problem. But it took her several days to build up the nerve she needed to ask the question. If no one had told what it was that crippled those brothers, she thought, their parents must have a good reason for not telling. So to give herself moral support she thought carefully about how to word her questions, and wrote them down. Then she called the boys' mother.

"I said, 'Maybe you wouldn't like to talk to me about it, but I sure would like to know what your sons have, because Betty's boy has it,' " Pearly recalls. "She told me that they had gone to the charity hospital in New Orleans with their boys and she said that the doctor said, 'We don't know exactly what it is, but there's a specialist, I think it was in New York.' And he told her that's what it was. And she said the name: 'Friedreich's ataxia, that's what he has.' And she said, 'I'll remember as long as I live, when I asked the doctor what caused that, and he says, that comes from venereal disease.' "

Pearly passed the information on to Betty, who was too preoccupied at the time with unraveling the tangle of her own genealogy to pay much attention. But Pearly continued to ask people in Morse to try to think of others who might have had the disease, and someone reminded her of a boy slightly older than Darren who always walked oddly. His parents would never talk about the matter, and everyone assumed the boy had drug problems.

She knew that her neighbor was the boy's uncle, so she called him and got the phone number of the boy's grandmother. Again, it was Friedreich's ataxia. "It all comes from that dirty sickness," the grandmother added. If Betty LeBlanc had suffered grief, other mothers were enduring shame as well, and needlessly. Syphilis is not involved in Friedreich's ataxia.

By the time Betty tracked down every strand of her ancestral ties to Derald, Pearly had discovered five families in the Morse area who were affected by Friedreich's ataxia, and Wilensky had given her several other leads to patients with Cajun-sounding surnames. She went back to Father Hébert's collection and began to track family names that were not her own, often on behalf of people who were not only unenthusiastic about her efforts, but categorically wanted not to know about them. Some members of her own family still do not.

Even when people did cooperate, it was tedious work. Many

records were handwritten, and she had constant eyestrain. So
many names had changed, from LeBlanc to White or LeJeune to
Young. The baby Euphemon turned out to be the same person
as the man Simon. But the same names kept recurring in the
background of families with ataxia: Aucoin, Blanchard, Breau,
Comeau, Dugas, Doucet, Duhon, Gaudet, Guidry, Hébert,
LeBlanc. . . .

Often Betty would work at the library into the evening and
then come home and work at the kitchen table late into the night.
She transferred the genealogy to scrolls; they stretched all the
way from one end of the kitchen to the living room fireplace,
nearly twenty feet. She knew this because her sons sometimes
unrolled them, to tease her.

She spent hours circling the table with pencils and a ruler,
tracing connections, creating an ever finer network of relationships. For all the families with Friedreich's ataxia, the trail eventually led back to Nova Scotia, where it stopped with two
common ancestors: Etienne and Antoine. A barrelmaker and his
brother, Father Hébert reported. They had come to Acadia in
1640, healthy but bearing the seeds of a tragic problem.

From time to time, Betty tried to explain to Darren what she was
doing, why the lines on the scrolls might be important to him.
But in truth, although she had an impressive stack of genealogies, she herself had very little idea what it all meant for her
children.

From time to time, she got very discouraged. Then she'd
call her mother.

"Don't give up," Pearly would say. "Don't ever give up.
We have to know we did everything we can for the children."

So when she heard that her cousin Clyde was traveling to
Canada to meet another Acadian genealogist, she pleaded with
him to ask if anyone knew anything about Friedreich's ataxia.
The Canadian contact, it turned out, had a name on the tip of
his tongue: André Barbeau, a neurologist at the Clinical Research Institute in Montreal, had traced the disease through
fourteen families of Acadian origin.

Immediately, Betty wrote to Barbeau. "When I found out
this was a genetic disease, I began a family research," she told
him. "I discovered my husband and I have the same grandpar-

ents six generations back. We are both from Acadian [descent]."
She went on to say that her mother had found four additional
Cajun families with Friedreich's ataxia. "I am trying to do their
family tree to see if they are related to us. I heard of eight families from Lafayette, Louisiana, with FA. . . . I feel sure we are
all related in one way or another. Our ancestors came mostly
from Nova Scotia and Canada."

Barbeau replied six weeks later, saying he was "extremely
interested." He urged her to contact him again when the genealogies were complete. Well before then, however, she found
just the expert she had been looking for, much closer to home,
and entirely by accident. Even the scientist in question calls it a
remarkable coincidence. Betty calls it a miracle.

The incident happened at the home of Betty's cousin, Evelyne Goller. Betty's cousin is a specialist in Acadian language
and culture at Louisiana State University in Baton Rouge. She
is deeply proud of her heritage, and also of her new house—a
replica of an eighteenth-century Cajun farmhouse, authentic in
great detail, which she designed herself.

The house is in Breaux Bridge near Baton Rouge, in the
heart of Cajun tourist country. Although set far back from the
road, it announces itself with a large wooden sign that reads
TERRE TOUJOURS ACADIENNE: land forever Acadian.

Not surprisingly, the sign attracts attention, and passersby
often stop to take photographs of the house. Evelyne, who is
not by nature gregarious, usually ignores them.

One spring weekend in 1982, Evelyne noticed someone
standing at the roadside and describing the house's features to
the people inside a parked car, pointing and using broad ges-
tures. For some reason Evelyne herself can't define, she waved
and invited the strangers onto her veranda.

Among the party were a woman with a soft Southern accent, Mary Kay Pelias, and Robert Elston, a tall man with a cultivated appearance and a British accent. They were on the way
to a famous Cajun restaurant, they explained; they had noticed
Evelyne's house and stopped.

Pelias remarked that she was fascinated by Cajun culture.
They spoke about the house, and then for a long time about
Cajun French, and about the Cajun radio network where Eve-

lyne worked. It came out that Elston spoke fluent French, because his mother had come from France.

Given all that, Evelyne was mildly surprised to learn, well into the conversation, that Elston and Pelias came from a different academic culture from her own. They were both scientists at LSU's medical center in New Orleans, working in the department that applies statistics to the study of hereditary diseases. Just to be friendly, they gave her the office phone number before they left.

The incident came immediately to mind a few weeks later, when Evelyne received an unexpected letter from Betty. They had lost contact several years earlier. During a recent visit, Pearly had told Evelyne about Betty's children, but it did not occur to Evelyne to connect Betty with the geneticists until her letter arrived.

"Dear Evelyne," Betty began. "Had a good visit with Mama and Dad last week. She gave me your [new] address. We recently took our children to an Ataxia Clinic in New Orleans. I told the doctor about the many families I have found with Friedreich's ataxia in Morse and the surroundings. He wants me to try and put all these families together (if possible) and send it to them. I have been extremely busy doing this. Most families will not cooperate."

After reading the letter, Evelyne called the number for Elston's department. Pelias took the call, and listened with interest to the information about Betty LeBlanc. Pelias was occupied at the time with another project, and had no time to focus on Friedreich's ataxia. She thanked Evelyne kindly for the call, and taped the message note with Evelyne's phone number on the wall near her desk, where it waited for the rest of the summer.

She called back the following November, and asked Evelyne to set up a meeting with her relatives. A new staff member had joined the department, someone who happened to have a special interest in ataxia.

A week before Christmas, Betty finally prepared to meet the specialist she had been looking for. She gathered a fair sample of her genealogy materials, including the sheets that showed her own children, and placed them in a small brown briefcase. She and Derald drove alone toward Morse, where they picked up her parents and his mother, and then proceeded to Breaux Bridge.

Betty was terribly ill at ease. None of this was of her own doing, and none of it made sense to her. An unknown group of doctors (or so she thought of them) wanted to examine her records—but not, strangely, her children with ataxia. Instead, they had asked to meet her parents and her mother-in-law, who were healthy. In Evelyne's living room, Betty sat and bided time, wondering.

It was nearly an hour before the strangers arrived. Evelyne opened the door to see two women: Pelias, who she already knew, and a stranger. Pelias, vivacious and brightly dressed, made introductions. Her companion was Bronya Keats, a fellow geneticist.

If Pelias is vivid, Keats is subdued. She dresses plainly and speaks seldom, always softly and plainly, and with an accent Betty took for British but which is really Australian. The Friedreich's ataxia project would be hers.

The two researchers asked if they might use a table, and Evelyne offered the dining room. Everyone crowded around the small table, and Pelias began asking questions, starting with Pearly Vincent. They were many of the same questions Betty had asked her relatives a year earlier, but Pelias also inquired about ailments and causes of death. As Pearly Vincent began to speak, Pelias unfurled a roll of butcher paper and began sketching a family tree on it. Back in the lab, the geneticists would refer to this by a term usually reserved for royalty and racehorses: pedigree.

After she had told the geneticists everything she knew about her own genealogy, Pearly turned to Keats, who was sitting quietly, taking notes. "What about this venereal disease?" she said. "Could Friedreich's ataxia ever be caused by venereal disease?"

Keats looked directly into her eyes. "No," she said earnestly. "Never. If only it could. Then we'd have something to go on."

Pelias moved ahead, questioning Betty's father and her mother-in-law. Betty's own turn came last. "Tell me about your children," Pelias began.

Instead of answering, Betty lifted the briefcase and spread out her own sheets of paper. She pointed to Darren's name, then Dena's and Dana's. Here was herself, and Derald, and here were her parents and his, and way up here—and there, and here, and there again—the links that joined her ancestors to her husband's. She pulled out another sheet, and showed how the names on it interlinked with those on her own.

"I don't know if this is really what you wanted," she said, riffling through the stack. "I might could have brought some more. I didn't know what to bring."

"You mean there's more?" asked Pelias.

"Oh, yes," Betty said. "There's four other families in my hometown, and I also know about fourteen families in Canada. You just can't believe how the families with Friedreich's ataxia are all the same people. They're all related."

Pelias looked across at her colleague and spoke—deeply, this time, and slowly: "We have found a gold mine."

For all the physicians who had examined Darren and everyone else with his disease, no one had come close to an explanation for it. The person who finally used Betty LeBlanc's information to move closer to an understanding of Friedreich's ataxia was, put simply, a mathematician.

Keats's expertise is to use statistics to analyze patterns of inheritance, in order to assess how various traits pass from one generation to another. Until she moved to New Orleans, Keats's research subjects had not been real people. They were statistics, numerical results of blood tests, sent to her in the mail.

In her previous job, at the University of Hawaii, she had analyzed test results from people with a different form of ataxia, one that is inherited as a *dominant* trait: a single gene from one parent is sufficient to pass the disease to a child. When she moved to New Orleans and then learned about Betty's family, Keats was eager for an opportunity to launch a study of a recessive form. She also looked forward to being part of the research process from beginning to end.

The gift that Betty LeBlanc gave Bronya Keats at their first meeting was a wealth of detail about a *kindred*: a very large, linked group of families with a clear tendency to develop one inherited disease. Especially in the case of a rare disease, a clear hereditary pattern in a large family offers medical researchers a path out of a maze that seems to be nothing but dead ends.

A suitable kindred could point researchers toward the gene or genes responsible for a disease, which in turn could be the first step toward a treatment. This had been true for only a few years at the time Bronya Keats met Betty LeBlanc. They found each other at the start of a new chapter in the history of human biology, as we humans were beginning to read our own genes.

Techniques in genetics that were barely dreamed of that evening would soon make much more of a pedigree than a mere paper trail. Researchers were beginning to follow that trail in the DNA itself, by detecting differences between the DNA of different people. With luck, the trail could lead to the faulty genes themselves. Suddenly, it was far more interesting than ever to follow genetic histories.

Within months after the meeting in that Louisiana dining room in 1982, researchers elsewhere would report locating the genes behind two other mysterious hereditary diseases: Huntington's disease and muscular dystrophy. A few years after that, the precise molecular details about the cause of the most common form of muscular dystrophy—the exact nature of a structural component missing from the muscle cells of affected children—would emerge from research that followed the discovery of the gene. If medical scientists did not yet have a solution for that disorder, at least they knew exactly what needed fixing.

Betty's kindred, Pelias explained, could be the start of a search for the location of a gene that causes Friedreich's ataxia. It would be a biochemical search for a needle in a haystack—a single, flawed gene among the 100,000 genes contained in any person's DNA—but it had succeeded before. All it would take, Pelias told Betty, was getting blood samples from many of the people on her sheets who were alive, and then trying to find a pattern in their blood. It would take that, and time.

It's just a matter of luck, she went on. With luck, we could find the gene in two years. It was the first time any expert said anything to Betty about Friedreich's ataxia that she could honestly interpret as hope.

A few days later, Mary Kay Pelias wrote to thank the LeBlancs for coming to Breaux Bridge. ''Bronya mentioned your correspondence with Dr. Barbeau,'' she wrote, ''and—wouldn't you know—Dr. Elston had collaborated with him on a project several years ago. Robert called André for a chat about our new project, and now we're trying to coordinate a huge field trip to the Lafayette area during the first week of February.''

The clinic took place as planned, on a rainy weekend in early February 1983, at a hospital in Lafayette. The impending arrival of experts from New Orleans and all the way from Montreal was big news for several local papers, so attendance was good. Barbeau and Wilensky both examined about twenty people to verify that they indeed had Friedreich's ataxia, and some seventy family members listened to an address by Barbeau.

The sight of a dozen-odd wheelchairs at the clinic was sobering to Darren. If nothing was done, he began to admit, he would die of Friedreich's ataxia.

At the clinic, for the first time, Darren met Bronya Keats. For the first time, also, he was favorably impressed by an expert. Keats spoke frankly and directly to him. She seemed to know what she was talking about, and to have a plan.

''I saw Dr. Keats,'' Darren said, ''and I thought, Finally! There's someone out there who's doing some work, instead of just telling me, 'Go home and don't worry about it.' That started to make me change inside. I think as long as you know there's something out there, that there's a chance, there's someone working, that's what makes you feel good.''

In a very short time after she met Betty and Derald, Keats found herself drawn into their family. She brought her own family, including her parents, to stay with them in Orange. She talked to Betty about genetics and Friedreich's ataxia, over and over. As Betty had, she became involved with the local chapter of the National Ataxia Foundation. And she spoke with Darren about his future.

''It was a little scary getting close to that family,'' she said later. ''I'd want to get a positive result. Before this, I wanted my research to lead to a publication. Once I was involved with the families, I wanted to find a linkage. I worried that we wouldn't find it, that the families would get disillusioned. I wanted to find it for them.''

In genetics, a *linkage* is a reliable correlation between the inheritance of a disease and some other hereditary trait. Keats would be looking for some *polymorphism*—some physical difference between individuals—that coincided with the occurrence of Friedreich's ataxia in Betty's kindred.

The word *polymorphism* comes from the Greek, and means "many forms." We recognize each other by visible hereditary polymorphisms: skin color, hair and eye color, body stature, nose shape, and so on. If everyone with Friedreich's ataxia had blue eyes, the association would be obvious. Keats had to find a less obvious link. She meant to begin by looking at a few proteins.

The bulk of the information in DNA is interpreted for the purpose of creating proteins, the molecules that make up most living structures such as bones and muscles and organs and also carry out the chemical processes of life such as the metabolism of food, the reactions of the senses, and the interactions of hormones. There are often subtle variations between proteins— blood types are a good example—and these differences are specified by variations in the corresponding genes.

Unlike the Denver specialist and others who had studied Friedreich's ataxia, Bronya Keats was not necessarily looking for a biochemical abnormality involved in causing the illness. Any inherited trait that merely *coincided* with Friedreich's ataxia would do.

If people in Betty's pedigrees who had to have at least one gene for Friedreich's ataxia—either affected people or their parents—almost always shared the same protein variation, such as a certain blood type, it would be a very strong clue to the location of the gene for the disease. It would imply that the genes for the two traits—ataxia and blood type—were located very close together in the chainlike molecules of DNA that contain hereditary information. The general location of the gene would be a first step to its discovery.

For a few researchers, this kind of study is made simpler because they can take advantage of the most common normal polymorphism: gender. Certain diseases, such as hemophilia and muscular dystrophy, are called "sex-linked," because they almost invariably affect males.

That fact gives scientists an invaluable clue about where to look for the gene behind such diseases, among the twenty-three pairs of human chromosomes, the large molecules that contain DNA. Twenty-two of the pairs are called *autosomes,* and appear to contain two identical chromosomes. The twenty-third pair, an X and its smaller mate, the Y, are visibly different under the microscope. They're called the sex chromosomes, because they determine the gender of individuals.

Girls, who have two X chromosomes, very seldom inherit sex-linked diseases because they possess a second and normal X chromosome to compensate for any defect on the other X. Boys, who inherit an X from their mothers and a Y from their fathers, are susceptible to sex-linked diseases caused by defective genes on the X chromosome, because they lack a second X.

The genes behind many other normal polymorphisms or variations, such as the ABO blood types, occur on the twenty-two pairs of nonsex chromosomes, the autosomes. In most cases, the members of a pair of autosomes cannot be distinguished from each other as the X and Y can.

Chemically stained and spread out on a microscope slide, the chromosomes look like tiny worms, or perhaps very small pairs of striped socks. A trained eye can pair them up, and distinguish the X and Y chromosomes from each other and from the various autosomes. But there is no way to see particular genes in such a microscopic image of the chromosomes, a *karyotype.* Genes are much too small. What we know about them, we learn by deduction.

If a genetic defect that causes a disease is located close to the gene that controls the innocuous polymorphism, such as the gene for the blood type or for a number of other detectable proteins, the disease gene is said to be "linked" to the normal variation. Close enough, and they tend to travel together generation after generation within a particular family, despite all the reshuffling of genes that normally takes place during reproduction.

It does not follow that the polymorphism is the cause of the disease, or that everyone with a certain variant of a protein linked to a disease will develop the disease, any more than everyone who is a male will develop muscular dystrophy or hemophilia. But if a certain pattern of polymorphisms in a certain kindred does coincide with the pattern of the disease itself, and

if you know the location of the gene behind the normal polymorphism, it can be a very strong clue to the location of the
disease gene itself.

Geneticists call such a coincidence of polymorphisms a genetic *marker*. It is analogous to other kinds of markers—the ancient points of known location, such as cathedrals or stones
carved with the distance to nearby towns, which have been used
time out of mind as reference points for maps of the landscape.
To geneticists wandering the chromosomes unaided by a good
map, genetic markers are molecular landmarks.

If we are lucky, Pelias had told Betty, we could have a
marker within two years. But it took two years simply to get
funding to start the study. Then she began to search for a link
between the pattern of Friedreich's ataxia in Betty's kindred and
the inheritance of polymorphisms for ten different major blood-
type variations and for another twenty proteins in the blood
whose genes had been identified.

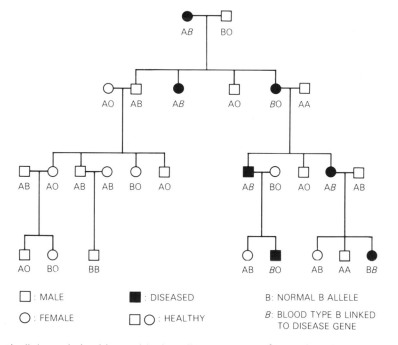

Genetic linkage I. In this mythical pedigree, a gene for an imaginary disease is
tightly linked to the gene conferring the blood type B on one particular chromosome. Note that people who have inherited a *different* chromosome containing
the B-type gene are not affected by the disease. The linkage occurs only in certain
individuals, but there is a clear pattern.

Bronya Keats was not lucky. She reported her results to the *American Journal of Human Genetics* fully five years later. By then, she and her co-workers had been able to do no more than rule out about 20 percent of the total area of the chromosomes as the location of the genetic defect behind Friedreich's ataxia.

It had become a collaborative effort by then, a combination of studies of blood drawn from Betty's kindred, from the Canadian families contacted by Barbeau, and from British families with Friedreich's ataxia being studied by another geneticist, Susan Chamberlain. Of the three groups, only the British geneticist had access to the newest tools for the search through the genetic haystack.

Chamberlain had the capability to look for polymorphisms in the genetic material itself—for small variations inside the DNA—not just for variations in the proteins, the molecules manufactured according to instructions in the DNA. By breaking up the DNA in special ways, scientists were beginning to find variations here and there that sometimes coincided with the presence of hereditary disease. These subtle differences in DNA were a new kind of genetic linkage.

It was Sue Chamberlain, not Bronya Keats, who eventually found the linkage for Betty's family, in early 1988, in blood from British kindreds. With help from Canada and Louisiana—building on the earlier information from Keats's first report on Friedreich's ataxia—she began to search through DNA from families with the disease, looking for fragments that definitely did not coincide with the disease. In this way, she was able to rule out most of the chromosomes.

Eventually, she was able to narrow the region of study to relatively small areas of three particular chromosomes. "Within a few weeks," she says, "we showed a linkage to chromosome 9."

Her report appeared in the British journal *Nature* on July 21, 1988, and pinpointed the gene. In families with the disease, affected people almost invariably possessed a certain variant of a DNA fragment known as MCT112. Until then, the fragment had no meaning whatever; a geneticist would call it "anonymous." It was merely a small stretch of DNA localized to a certain spot among the human chromosomes, a fragment that a researcher somewhere in Utah had found to occur in more than one form.

Of all the fragments Sue Chamberlain tested, MCT112 happened to show the pattern she was looking for. It matched the inheritance of Friedreich's ataxia. The few scientists searching for an understanding of the disease had their first landmark.

In the end, Betty's tireless research had helped significantly, but only to identify where the gene is not. Bronya Keats's name appears in the Chamberlain report only as an acknowledgment at the end, among some twenty research groups that contributed to the advance.

Nonetheless, Keats was jubilant. As soon as the report was public, she began to send letters to everyone in Betty's kindred who had given blood. "This breakthrough means that we can concentrate our efforts on this particular chromosome, rather than continuing to search on all the chromosomes," she wrote. "We hope that within the next year we will have a blood test that will allow us to determine very precisely which members of your family are likely to be carriers of the gene for Friedreich's ataxia."

What's more, it allows scientists for the first time to decide definitively whether Friedreich's ataxia has several variants or only one, and to decide who really has this neuromuscular syndrome as opposed to one of the others.

The first person Bronya notified, of course, was Betty, who mistook her jubilation to mean that the British researchers had found the gene itself, not just its approximate location as identified by a marker. To Bronya, the finding is a watershed in Friedreich's ataxia research; to Betty, once she understood what had really been found, it was almost trivial. A staunch Catholic, like most Cajuns, she has little interest in a search for carriers of the gene. The result of discovering that two parents are carriers is often prenatal detection of an affected fetus; the result of that is often abortion.

"Step one has no interest to me," she said. "They had to find that to go further. They will find the defective gene. That's our goal. Every time they find a marker, they're getting closer."

After completing the pedigrees, Betty had been busy for several years cajoling people in Cajun families to give blood for genetic studies, and arranging for the blood to be drawn and delivered to Keats. The latest research from England made it obvious that this was no longer critical. Her work in that sense was done; all she could do now was wait.

Early death, that's for someone else's children, she told herself. If we work hard enough, we'll find a cure. We have to.

A year after Chamberlain's report appeared in *Nature,* only one more genetic marker for Friedreich's ataxia had been discovered, but it pinned the location of the gene conclusively. Statistical studies showed astronomical odds—10^{50} to one—that the gene was very close to the two markers. As with muscular dystrophy, finding the actual gene held the prospect of understanding, at last, what was wrong with Darren LeBlanc.

When the first marker for the Friedreich's ataxia gene was isolated, Darren was entering his senior year as a biology student at Lamar University in Beaumont, Texas. It was an ideal campus for a young man rapidly losing his mobility: small, flat as an industrial park, quiet.

During college Darren began to use a cane and, eventually, a walker, and by graduation he needed a motorized cart to get around the campus. In his one-room apartment, desirably small for someone with his problem, Darren moved about by shifting his weight from one piece of furniture to another. He cooked for himself, and did all his writing on a computer.

"I never think that I'm going to get worse, but I do," he said a few months before his graduation, still sounding as if he could not assimilate the fact. "When I started college, I was thinking, when I'm a senior, I'll still be walking. I never think that I'm going to slowly progress. I think this is the worst I can get, I can walk with the walker and I'm fine. I don't think I'll ever lose the ability to walk with the walker but I guess one day I will. . . . It's like, it doesn't cross my mind to think that way."

Darren intended to continue into graduate school and to become a neuroscientist, to study the brain, with a particular interest in his own illness. Quite aside from his handicap, he had shown unusual initiative as an undergraduate: his independent research on two food additives was good enough to be accepted and published in a minor scientific journal.

On his applications to three graduate programs, Darren revealed his illness because he felt it would be unwise to surprise anyone who might invite him for an interview. But his writing was so slow and imprecise that he did poorly on the graduate qualifying exams. Stubbornly, he at first refused to ask for any more time on account of his disability. But later he relented, took the test again, and the scores were satisfactory. However, the record showed that he had taken the test under special conditions. No graduate school would invest in Darren LeBlanc.

Nonetheless, he hoped to be allowed to contribute somehow to Friedreich's ataxia research. He decided to study for a second bachelor's degree, in biochemistry.

"I feel that there's so much we really don't know, and that's what I like about it. There's so much to be discovered. . . . I also feel that I have so much in common with that field, and I really feel that I would be so interested in it. There's no cure, there's no treatment. I just feel that there's got to be something more than that."

It was the same thought he had had years earlier, every time he left a specialist's office. By now, he understood that everything his mother had done was her own response to the same thought, and that she had been trying to help him. It baffled him that he was not being allowed to help as well.

"I don't understand how some people could think that only physically abled scientists are able to solve the world's problems," he wrote in a letter several months later. "It does not matter who is solving a problem; what is important is how the answer to that problem will benefit mankind."

The same Darren LeBlanc once stared at photographs at the United Nations and resolved to help suffering children. He still means to do that, given a chance.

2

GENETIC DETECTIVES

We leave Darren in limbo, his life story incomplete, the story in his genes an enigma. But there are thousands of other stories, perhaps one for every single human gene. Some of them are coming, very suddenly, to a resolution.

While Darren LeBlanc was sweating out both a hot Texas summer and the fate of his graduate school applications, Zoe Weaver completed her junior year at college and went home to New York City, trying hard not to be more jubilant than she knew she ought to be. Zoe also has a genetic disease, but unlike Darren she was allowed to be part of a detective team that was on the trail of the killer that was after her: the cause of cystic fibrosis. By the summer of 1989, they had it cornered.

Zoe was studying microbiology at the University of Michigan and, like Darren, acting on the premise that the error in her genes would not interfere with a research career. Zoe also has an unexplained recessive disease, but hers is a very common and, if anything, a more severe one than Friedreich's ataxia. Cystic fibrosis affects one in twenty-five hundred Caucasian

American children. It causes mucus to accumulate until it clogs small passages in the lungs and other organs. About half of children born with CF do not survive to Zoe's age. The disease is fatal.

During a routine medical checkup that spring, a social worker had asked Zoe if she would be willing to give a blood sample for cystic fibrosis research. "Okay," she said, "but I'd like to know who's doing it, and what it's for." The social worker put her in touch with Dr. Michael Iannuzzi, who was part of a large team of researchers at the University of Michigan looking for the gene behind cystic fibrosis.

When Iannuzzi met Zoe and learned her major, he offered her a job. After classes ended in May, she began working in his lab full time, analyzing blood samples like her own. Again and again, she had to do the molecular equivalent of a fingerprinting: cultivating suspect DNA fragments inside colonies of bacteria and testing them against other bits of DNA, looking for a match.

The search had been narrowed to chromosome 7 several years earlier. Now a number of teams were racing to locate the gene precisely.

"There's a reasonable likelihood that the cystic fibrosis gene now exists in a tube," the lead scientist on the Michigan team, Francis Collins, had remarked at a genetics conference the previous autumn. By the time Zoe joined the effort, he thought they had found it. Iannuzzi's lab was busy testing it against DNA from the tissues primarily affected by cystic fibrosis, the lung and the pancreas. Together with collaborators Jack Riordan and Lap-Chee Tsui in Toronto, they were trying to confirm a difference between normal DNA and what they thought was part of the CF gene.

"It was very low-key," Zoe recalls. "Everyone wanted to be very calm about it. I guess everyone was real cautious." A few years earlier, an English scientist announced he had found the CF gene, and then had to retract his words. No one wanted a repetition of that.

This was one of many reasons that, during a lab meeting in mid-July, Collins asked everyone to keep the good news a secret. He was preparing a result for publication. The Michigan and Toronto teams had found a very tiny deletion—a small segment missing from DNA samples—in 70 percent of CF patients. The same segment was present in every normal sample. "At that point, I felt excited," Zoe said. "But it wasn't an all-of-a-sudden, Eureka! sort of thing. You think you may have found it, but you're not sure. Finally you accumulate enough evidence that you go ahead."

Zoe went home to her family for the month of August. Near the end of the month she saw a report in the newspaper, and then on television, that the cystic fibrosis gene had been found. Zoe was astounded. She called the lab in Michigan, to ask what had happened to justify the breach of security. Nothing except a leak, she was told.

Immediately, the leaders of the Michigan and Toronto teams had decided to hold a press conference, in order to keep things from getting out of hand. "It was real important to everybody that people not get the impression that we had found a cure, that people not get their hopes up," Zoe explains. "They wanted everyone to have the impression that there was still a lot of work to be done."

Without pausing, she takes the other side. "I mean, there should be a lot of hope. There's certainly nothing wrong with excitement."

Many Americans will remember August 24, 1989; it was the day Pete Rose was banished from baseball. For that reason, most of the footage from the euphoric press conference about cystic fibrosis never made the evening news. Few people saw the researchers' beaming faces, announcing they had found the CF gene. It was the kind of event Betty LeBlanc dreams about.

Robert Dresing, president of the Cystic Fibrosis Foundation, said he has been listening to the story of cystic fibrosis ever since it was introduced into his life—his twenty-two-year-old son has CF—and that the story "has never been anything but despair."

Now this is no longer the case, he proclaimed. "We know we can change the face of things. We know that we can develop the technology and create the science. We have that. The only thing we don't have is the questions that haven't been answered yet. But we will have those answers."

The CF discovery was momentous. Some 4 percent of Americans carry one copy of the CF gene; now a screening test could be developed, and the prospects for a new treatment were suddenly very bright. But the significance of the discovery was greater than that.

Dresing's jubilant words could apply equally well to any other disease, or in fact to any inherited facet of human nature. We know we can develop the science and the technology to understand it now, because finally we have the means to find a gene, any gene at all. The only thing we don't have yet is all the questions. But we know we will have the answers.

A major project supported by the federal government has the stated goal of discovering the exact contents of our DNA, the human genetic material, within about fifteen years. The techniques exist to learn the precise contents of human DNA, and the urge to do so has become overwhelming. For a decade, research has been proceeding in that direction, almost inexorably.

At the outset, researchers of the human DNA gained most from studying diseases. Hereditary disorders are the most obvious and most compelling subjects for genetic research, and in turn medical research has gained a great deal from the new tools of molecular genetics.

The work of the next fifteen years—first creating a crude map of various regions of the genetic material, and later a detailed molecular atlas—is only a technical chore that scientists are hurrying to complete. For them, the real excitement lies beyond, when they can embark on an exploration of our very biological essence, using the atlas. Along the way, countless Zoe Weavers will have their own reasons to celebrate.

The discovery of the cystic fibrosis gene was especially momentous to geneticists, because it proved that they could discover the tiniest possible genetic flaw, without any outward clue to its nature, by combining a number of their newest techniques. Having done so, they were on the trail of a new protein that promised to explain what goes wrong in the disease.

It must have been like solving a perfect crime years after the fact. At the press conference, Francis Collins of the University of Michigan was less romantic about it, but still elated. He described the advance as taking researchers beyond a tremendous bottleneck in the understanding of the disease. Scientists could now move forward, he said, to make rational decisions about research and medical care. "It is a milestone of the highest order," he said.

Cystic fibrosis is the most common severe autosomal recessive disease among Caucasian Americans. Because it is a great masquerader—often mistaken for malnutrition, pneumonia, or bronchitis—CF was not recognized as a distinct syndrome until about fifty years ago. The abnormally thick, sticky mucus characteristic of CF becomes infected easily and eventually destroys the lungs.

"The question always posed is, Why do these individuals produce abnormal mucus?" says Robert Beall, medical director of the CF Foundation. "What is it about their genetic makeup that causes them to produce this mucus? If we really wanted to identify new and effective therapies, clearly we had to identify the gene."

The $15 million the foundation spent chasing the gene was "the biggest bargain in our lives," he said afterward. Beall estimates that the total annual medical bill for Americans with cystic fibrosis is something like $300 million.

The process that revealed the defect behind cystic fibrosis (the same general method Bronya Keats and Susan Chamberlain were using at the same time) is known as *reverse genetics,* because the only previous way to find a gene had been the other way around. By definition, going "forward" had meant starting with the discovery of an abnormal protein and then searching for the gene that encoded it.

In fact, "reverse genetics" is a misnomer. Nature usually works the other way around, from genes to proteins. The discoveries that led to the genetic revolution also began with studies of heredity and proceeded to DNA and to the breaking of the genetic code. That studies of proteins went on all along was due to the simple fact that many of them are easy to extract and study.

The Austrian monk Gregor Mendel deduced the basic laws of heredity in the 1860s, by studying the breeding of plants and flowers. Long afterward, the molecular basis of heredity remained a total mystery. It took many decades for people to become comfortable with the idea that the basis of life was molecular in the first place, and decades more to isolate and understand what Mendel called the "factors of heredity."

For generations, the issue of finding those "factors" seemed too complicated to resolve. When James Watson and Francis Crick did resolve it, their answer was astoundingly simple. In 1953, they figured out how DNA—a molecule that until then had seemed as structurally interesting as a railroad track—was in fact the molecular message of heredity, complex and reproducible. Other scientists, afterward, worked out the code that carries the message.

It is a remarkably simple code: a series of chemical symbols that, in different combinations, translate into the twenty basic units that make up proteins. Cells in the human body are equipped to interpret that code, to read it and convert a string of DNA subunits, or *nucleotides*, into another sequence of molecules, the amino acids of which proteins are composed.

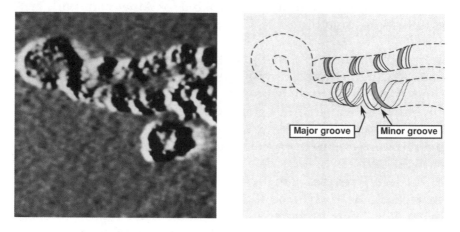

DNA: scanning tunneling micrograph and explanatory diagram. Courtesy of Lawrence Livermore National Laboratory.

The code in DNA, the string of chemical letters, is made of only four symbols: the nucleotide molecules called *adenine* (usually abbreviated as A), *cytosine* (C), *guanine* (G), and *thymine* (T). Like the code an espionage agent would carry, the DNA code is a long and (to an untrained eye) monotonous, unbroken sequence of these subunits, like this: ATTTCGTTCTG-TAATTTTGTC. . . .

Organisms can reproduce, and cells can function, because they are able to copy DNA precisely. This is due to a chemical property of the four nucleotides. They bind to each other chemically, but only in a certain way: A binds only to T, and C binds only to G. Cells use this property of complementarity to translate the code, and biologists also use it to study DNA.

The intact DNA molecule is not a single strand, but a double one, a matched pair of DNA sequences. It's rather like a zipper, except that there are four kinds of teeth, and it will not close unless all the teeth match properly. The DNA zipper is twisted and twisted again. In nature, DNA is a double helix.

During reproduction, the zipper unwinds and opens, and the nucleotide "teeth" from each half are copied in complementary fashion, an A for a T and a C for a G, and so on. An old half-zipper gets a new, complementary other half. The genetic information in a cell can be reproduced indefinitely and—barring accidents—precisely.

The accidents are known as mutations. There are two kinds: those caused by external insults, such as radiation or chemical mutants, and those caused by mistakes in the genetic process itself. The natural editing errors of the reproductive process include random breaks, deletions, and mismatches in DNA that periodically occur during cell division or reproduction. Many of them occur during the normal reshuffling processes of reproduction that make every individual unique.

Most mutations are harmless; some are catastrophic. If a mistake happens during the creation of sperm or eggs, it may cause a hereditary illness. If it happens in other kinds of cells, it will affect only that individual, but it still may cause a disease such as cancer.

Written as an unbroken sequence of alphabetical symbols for nucleotides, the genetic instructions to produce a human being

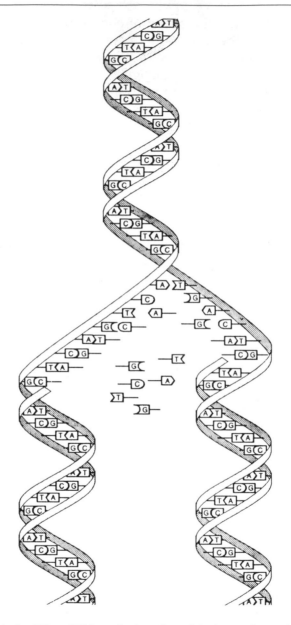

The double helix. When DNA replicates, the original strands unwind and serve as templates for the building of new, complementary strands. The daughter molecules are exact copies of the parent, each daughter having one of the parent strands.

SOURCE: Office of Technology Assessment, 1988. Reprinted from *Mapping Our Genes—The Genome Projects: How Big, How Fast?* OTA-BA-373 (Washington, D.C.: U.S. Government Printing Office, April 1988).

would fill 5,000 books the size of this one. They are contained in the chromosomes, a set of filamentous polymers altogether more than two yards long but infinitesimally thin, and compacted 100,000-fold until they would fit into a space smaller than the period at the end of this sentence.

This book is written using an alphabet of twenty-six symbols assorted into words of five or ten symbols; in the same way, genes are different assortments of the same four symbols, the nucleotides, but they are words thousands of symbols long. Today the term *gene* is far more than a loose concept; it is defined as a specific biochemical unit of genetic information, one word in a very long molecular document. There appear to be about one hundred thousand human genes; each one specifies the ingredients of a particular protein.

Figuring out how it all works is the task that remains for scientists in the next century. Genetic researchers have already worked out the DNA sequences that signify the beginnings and

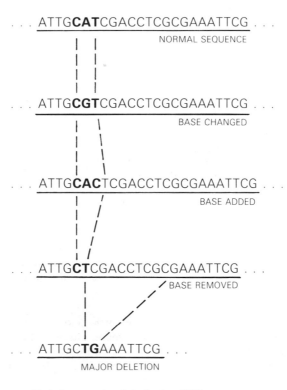

Mutations: misprints in the DNA message

endings of genes, as well as stretches of DNA that are common between related groups of genes, and the identities of some large sequences that appear to be meaningless.

In the 1960s, Marshall Nirenberg at the National Institutes of Health worked out how the DNA sequence translates into proteins. Set out on a chart, the genetic code is as simple to interpret as the mileage grid at the back of a road atlas. Different sets of three DNA nucleotides correspond with different amino acid components of protein.

TABLE I   THE GENETIC CODE

	A		G		T		C		
A	AAA	Phe	AGA	Ser	ATA	Tyr	ACA	Cys	A
	AAG	Phe	AGG	Ser	ATG	Tyr	ACG	Cys	G
	AAT	Leu	AGT	Ser	ATT	Stop	ACT	Stop	T
	AAC	Leu	AGC	Ser	ATC	Stop	ACC	Trp	C
G	GAA	Leu	GGA	Pro	GTA	His	GCA	Arg	A
	GAG	Leu	GGG	Pro	GTG	His	GCG	Arg	G
	GAT	Leu	GGT	Pro	GTT	Gln	GCT	Arg	T
	GAC	Leu	GGC	Pro	GTC	Gln	GCC	Arg	C
T	TAA	Ile	TGA	Thr	TTA	Asn	TCA	Ser	A
	TAG	Ile	TGG	Thr	TTG	Asn	TCG	Ser	G
	TAT	Ile	TGT	Thr	TTT	Lys	TCT	Arg	T
	TAC	Met	TGC	Thr	TTC	Lys	TCC	Arg	C
C	CAA	Val	CGA	Ala	CTA	Asp	CCA	Gly	A
	CAG	Val	CGG	Ala	CTG	Asp	CCG	Gly	G
	CAT	Val	CGT	Ala	CTT	Glu	CCT	Gly	T
	CAC	Val	CGC	Ala	CTC	Glu	CCC	Gly	C

The genetic code: Each three-letter sequence composed of A, T, C, and/or G stands for a *codon,* a sequence of three DNA nucleotides. Each codon is followed by an abbreviation for the amino acid, or protein building block, it specifies. Cells read the code to determine the order of amino acids when assembling proteins.

Ala = alanine; Arg = arginine; Asn = aspartic acid; Cys = cysteine; Gln = glutamine; Glu = glutamic acid; Gly = glycine; His = histidine; Ile = isoleucine; Leu = leucine; Lys = lysine; Met = methionine; Phe = phenylalanine; Pro = proline; Ser = serine; Thr = threonine; Trp = tryptophan; Tyr = tyrosine; Val = valine.

■   ■   ■   ■

So far, less than about one-thousandth of the total human DNA has been sequenced, its precise ACTG sequence deduced. Trying to understand it purely by learning the sequence, 3 billion nucleotides one after another, would be a mammoth and unrewarding task. (Chemical methods for sequencing DNA are automated today but still relatively slow, given the total amount of DNA in the human genome.)

It's not obvious what meaning anyone could derive from that long sequence of symbols anyway. Making sense of it would be worse than trying to read a book printed without spaces between the words.

What's a gene, and what's garble? Where in the sequence should you start marking off groups of three? Biologists know parts of the answers to these questions; still, analyzing that 3-billion-letter sequence would be highly arduous.

The more productive approach that has been under way for nearly a decade is reverse genetics, trying to locate individual genes and learn what we can from them. This began with observations about the heredity of certain proteins, and their relation to inherited disease.

Researchers began by looking for genetic links between hereditary diseases and various well-understood medical traits such as blood types or the tissue types used in organ transplantation. In some cases, the genes that encoded these characteristics had already been found.

It was logical to wonder whether a trait that coincided with a hereditary disease did so because the gene behind that trait was physically linked to the gene that conferred the disease. Presumably, if they traveled together generation after generation, in a pattern so predictable you could follow it on a family tree, a trait and a hereditary disease must be encoded by genes that were adjacent on the same chromosome, or nearly so. Finding such a linkage to a marker is a critical first step toward pinning down the gene itself.

Cystic fibrosis was first defined in 1938. For years afterward, scientists searched in vain for any conclusive clue to the cause of the abnormal mucus. Theories abounded: CF was a disorder of metabolism, perhaps, or a flaw in the hairlike cells called cilia that should move mucus out of the lungs and the trachea.

It was not until 1983 that a substantial biochemical clue emerged: abnormalities in the secretory ducts of surface tissues and sweat glands that transport water, salt, and other ions between the blood, the skin, and the outside world. The reason for the defect was not obvious.

The first lead to the solution came from a genetic marker. In August 1985, scientists reported that CF families showed a genetic linkage to a certain protein. The location of the gene that encoded that protein was not known, and the protein itself was not a self-evident explanation for the disease. But the nature of the genetic linkage did suggest that a single gene was responsible for CF.

In the history of genetics, there had never been a more perfect time to begin searching for single genes. New molecular techniques made it suddenly much easier to pinpoint genes to

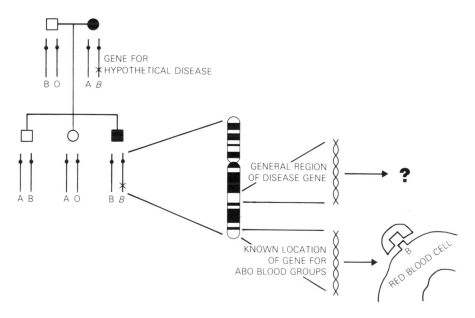

Genetic linkage II. The location of the known gene for the blood type is a clue to the identity of the unknown gene behind the disease. How often the blood type and the disease coincide is a clue to the distance between those two genes on the same chromosome.

individual chromosomes. Given the prevalence of cystic fibrosis, and the likelihood that a single gene was responsible, it was probably inevitable that the search for the CF gene became a race.

The first to pinpoint the CF gene to a particular chromosome were Lap-Chee Tsui of the Hospital for Sick Children in Toronto, and his co-workers at a Massachusetts biotechnology company, Collaborative Research, Inc. They tracked it down by finding a marker that was not a variant in protein but a detectable difference in the DNA itself. They accomplished this by searching through large collections of human DNA fragments for one that "segregated" with, or accompanied, CF through a pedigree. It turned up in mid-1985, a fragment extracted from chromosome 7. Tsui had the approximate address of the CF gene.

"Looking for a cystic fibrosis gene was like looking for a house in a city between New York and Los Angeles without a street address," says Tsui. Discovering the gene's general location on chromosome 7 was like narrowing the search to Chicago.

Within months after that, the search was narrowed to a smaller region of the chromosome, as scientists elsewhere found linkages to genetic markers that appeared to be somewhat closer to the gene. It was like narrowing the search to a single street in Chicago, but a very long one. The region between the two new markers was estimated at about 2 million nucleotides in length.

For most of medical history, the only way to study a disease was by examining something that could be extracted easily from the body: blood, urine, saliva, possibly skin. To explain a disease, there had to be some observable difference in a fluid or tissue that distinguished ill people from healthy people. It also had to be something relevant, not just coincidental. A scientist might toil heroically over something that was a by-product, not the cause, of a disease.

Sometimes medical researchers did find an answer to a familial disease without any recourse to genetics at all. Diabetes is a good example. The explanation followed an observation about that most easily extracted body fluid, urine.

Centuries ago, urine from certain people was known to at-

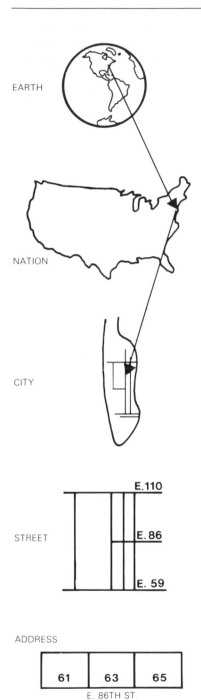

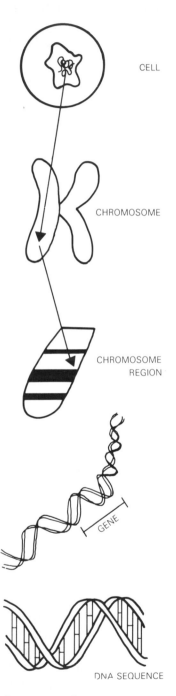

EARTH

CELL

NATION

CHROMOSOME

CITY

CHROMOSOME
REGION

STREET

E.110

E. 86

E. 59

GENE

ADDRESS

| 61 | 63 | 65 |

E. 86TH ST.

DNA SEQUENCE

Comparative scales: earth mapping and gene mapping.

tract flies. In 1815, a French chemist learned why: it contains a simple sugar, glucose. This led to the intuition that people who had "sweet urine" could not digest sugar properly, which led in turn to successful attempts to mimic the disorder in animals and eventually, in 1921, to the discovery of the protein deficient in diabetes: insulin. This discovery was substantiated, first in laboratory animals and later in humans, by injecting insulin into the blood and then measuring sugar in the blood or urine.

Polio, influenza, syphilis, and other infectious diseases, even AIDS, were susceptible to medical research because the microorganisms that cause them could be extracted from blood or infected tissues and cultivated in the laboratory. Other diseases, such as the hereditary bleeder's disease called hemophilia and many immune disorders, were traced to the absence of some factor present in normal blood. Always, it was possible to try to understand what went wrong in a disease if someone was lucky enough to detect an abnormality of some substance in a body fluid or tissue.

The roots of many illnesses, however, are sequestered in the internal organs, the muscles or nerves, or the brain, or they arise before birth. Or, like cancer, they develop over the course of decades. Far more often than not—for more than nine out of ten hereditary diseases—there has been no consensus whatever about the nature of a basic defect, lacking the gene that caused it. Any avenue of research was likely to be a dead end.

This was the predicament facing Dr. Donald Wood in 1980. An assistant professor of neurology at Columbia University, he had been looking for the cause of muscular dystrophy, a genetic disease that cripples about one of every 3,500 boys under the age of five.

Wood had found that muscle tissue taken from boys with the disease could not generate the normal amount of force. That was where he stalled. Beyond that, all anyone knew for sure was that eventually the muscle tissue disappeared completely. No one's biochemical explanation for the disease, including his own, stood up to close scrutiny.

"I could make inferences forever," he says. "It was frustrating. I had no dearth of ideas to pursue, but any alley was likely to be blind."

Wood was stymied. He left the laboratory completely, to

become research director of the Muscular Dystrophy Association. Soon thereafter a geneticist laid out for him the implications of an obscure, theoretical report in a scientific journal. The article explained how to find clues to the location of human genes by breaking DNA up with certain enzymes and testing the fragments in affected families.

It was the same method Lap-Chee Tsui later used to find the approximate location of the gene for cystic fibrosis. Once Wood heard about this idea, he says, "it was impossible to believe that is not the direction to go"—to locate the faulty gene common to children with muscular dystrophy, and proceed from there to find the gene's product, the defective protein.

This was especially true for muscular dystrophy, because researchers had a head start. They could eliminate 45 of the 46 chromosomes from consideration, because MD was one of the sex-linked diseases. Everybody knew the gene was somewhere on the X chromosome.

At the time there wasn't a single report of the disease in a female. Any girl who inherited a copy of the faulty gene for muscular dystrophy presumably had a normal copy of the gene on her other X chromosome, and would be healthy. But a boy could inherit the disease from his mother, because he inherited an unbalanced X chromosome. The Y chromosome he inherited from his father, the X chromosome's mate, presumably did not contain the normal gene. He would learn to walk, and soon after would find himself unable even to crawl.

The Muscular Dystrophy Association sought advice about the new genetic strategy from James Watson. The man who shared the Nobel Prize for the discovery of DNA's significance now headed a world-famous genetics laboratory at Cold Spring Harbor on Long Island. Watson was decisive: If you spend $5 million on studying DNA fragments as outlined in that article, he predicted, you'll have the gene for MD in three to five years.

The association took the challenge. It curtailed most of its support for biochemical fishing expeditions and devoted millions to genetics, gambling that someone could find the gene and, from it, the cause of muscular dystrophy. What's more, it chose to stake the money on a few good geneticists who were willing to suspend all other research until one of them found the MD gene. Among them was a Boston researcher named

Louis Kunkel, who until then had been studying the pure biology of the sex chromosomes. For Kunkel, at the outset, MD represented merely a good excuse to study the function of the X chromosome.

By that time, previous DNA studies had narrowed the region of the X chromosome under consideration, much as Bronya Keats's studies had narrowed the portion of the genome that Sue Chamberlain needed to search for the Friedreich's ataxia linkage. Louis Kunkel built his own research on studies of genetic material from the blood cells of one Bruce Bryer, a foster child in Spokane, Washington.

Bruce Bryer could neither walk, see, nor fight infections. He had been born with a trio of X-linked diseases: muscular dystrophy, a degenerative eye disorder called retinitis pigmentosa, and a rare immune-system defect known as chronic granulomatous disease.

Bruce Bryer's X chromosome was missing a large chunk of DNA. Kunkel began with the obvious assumption that whatever caused muscular dystrophy was the absence of something essential in that missing DNA. To close in on the segment involved in MD, he sought help from more than seventy collaborators around the world. Kunkel amassed DNA from more than three thousand other boys with MD, which he matched against Bruce Bryer's DNA. Often, the other boys' DNA showed smaller deletions within the region of DNA that was totally absent from Bruce's X chromosome.

With that as a lead, Kunkel's team proceeded to focus more closely on those smaller regions, comparing the genetic material from the X chromosomes of healthy people with that from boys who had muscular dystrophy. (At the same time, also in part by using Bruce's DNA, Kunkel and another research team located the gene for chronic granulomatous disease, with similar methods.)

After Kunkel and his massive team located the gene for MD, revelations about muscular dystrophy began to flow from the discovery with gratifying speed. A year later, Kunkel's team reported isolating the gene responsible for MD, and determining its sequence. Six months after that, they knew that a nearly identical gene occurs in mice, and that mice with a muscular weakness similar to MD lack that normal gene. By spring of

1988, they had extracted the defective protein itself from boys with muscular dystrophy. They named it dystrophin. The following January, the Kunkel team reported correcting the mouse version of muscular dystrophy with injections of dystrophin.

If no one could say how soon all this might lead to a cure for humans with the disease, at least there was finally an explanation for the baffling disorder that seemed to eradicate muscles.

Dystrophin, a rodlike protein, appears to be a structural support in the membranes of muscles, like the girders that buttress the walls of a skyscraper. Without it, as Wood had found years earlier, large muscles have no force and cannot support the skeleton. They collapse and atrophy. The body clears the dead muscle cells away, and they vanish.

"The work of Kunkel and his associates establishes without any doubt the power of the theory . . . that provided the basis for reverse genetics," crowed an editorial in the *New England Journal of Medicine* that accompanied the team's report of the newfound protein.

What made the subsequent cystic fibrosis discovery even more exciting than Kunkel's isolation of the MD gene was the fact that Tsui and his co-workers had found the CF gene without any initial hints whatever about its location. They had no Bruce Bryer to help them, no patients with cystic fibrosis who helpfully possessed a major deletion of DNA in the region of the gene. What they ultimately found, in about 70 percent of the blood samples from CF patients they tested, was a deletion of a single codon, only three nucleotides, coding for a single amino acid in the corresponding protein.

"This is the first time a gene responsible for human disease has been [isolated] solely on the basis of its chromosomal location, without the existence of a large rearrangement," says Francis Collins. "This success gives a real sense of hope to those people cloning genes for another four thousand genetic diseases, many of which can only be approached in this way. This is a successful strategy. It can work. It provides hope for a number of other diseases."

Having narrowed their search, figuratively, to that long street in Chicago, the cystic fibrosis researchers spent four years walking up and down, trying to find the right house. To the scientists and to patients—who did not know how long the search would take—it seemed interminable.

The 2 million nucleotides in that span of chromosome 7 were too many to analyze by DNA sequencing. Theoretically it would be feasible to get the complete DNA sequence, base by base, by making many thousands of clones of tiny DNA fragments and testing them on an automated sequencing machine. Even with the clones in hand, it would take roughly two hundred days, with the machine running flat out, to produce an unbroken sequence that would even then be meaningless. Instead, the researchers broke the region into fragments and began to analyze it piece by piece. It was research in the truest sense, back to the Latin origins of the word: circling it again and again, scrutinizing it from every direction.

In Toronto, Tsui began to "walk" the chromosome, making very small DNA fragments to analyze how they overlapped, in much the same way a geographer might map a mountainous region by matching overlapping aerial photographs. Ray White at the University of Utah and a British geneticist, Robert Williamson, had each found markers closer to the gene than Tsui's first one; using them as landmarks, Tsui intended to keep walking until he found the gene, trapping it inside ever smaller overlapping fragments.

Simultaneously, the Michigan team used Collins's strategy of gene "jumping" to map relatively large fragments of the region. By making circles of large DNA fragments and then studying the joined ends, it was often possible to move quickly across a region, ignoring large spans of irrelevant DNA in between. Eventually, he would have the gene surrounded.

Meanwhile, both teams looked for the fingerprints of the culprit at the scene of the crime—for signs of some genetic material corresponding to part of the region they were focusing on, but present outside the nucleus of a cell, in locations where genes are processed into proteins so they can carry out their functions. Somewhere, in a tissue relevant to CF such as the lung or the pancreas, must be a span of genetic material waiting to be processed into a protein that caused abnormal mucus. If they found genetic material corresponding to a region near the marker in some cell involved in the disease in the part of the cell where genes are processed into protein, it would be substantial evidence that they had tracked down the gene they were after.

Ultimately, they found only one fragment that matched part of the region of chromosome 7 implicated in CF. It turned up in a colony of cells cultured from human sweat glands by Dr. Jack Riordan at the Hospital for Sick Children. Even that result was rather equivocal. "At first we said, Nahhhhh," Collins recalls, "but then we looked closer."

Zoe joined the lab at that point. Later studies showed that a larger stretch of DNA containing that particular fragment also occurred in tissue from the lung, pancreas, and other organs involved in cystic fibrosis. Testing the larger fragment of DNA extracted from the region thought to contain the faulty CF gene, they found that nearly 70 percent of patients with the disease showed the same three-nucleotide deletion. None of 198 healthy people did.

The team did not announce a discovery until they had translated the DNA from the region into the corresponding amino acids, using the genetic code, and compared the result with the sequences of known proteins, to see if it was similar to any normal substances that might explain the physiological defects seen in CF. In September 1989, in the journal *Science*, they described the protein predicted by the translation effort. They had already begun to search for it.

The CF protein is structurally similar to a large class of some twenty proteins that are known to be embedded in cell membranes and that are involved in transferring substances across the membrane. Judging from these other molecules, the CF protein itself appears to span the cell membrane, and to be constructed in a way that allows it to capture energy from another kind of molecule and translate it into hydraulics, which allows it to shuttle substances across the membrane. The single amino acid that is missing in CF occurs in the general location at which the protein would capture the molecular energy necessary to act as a shuttle.

"We think—although it has yet to be proven—that this . . . might cause a water transportation problem in the cell," Riordan explained at the press conference, "causing the CF patients to suffer from thick, sticky mucus. . . . As a result, the mucus blocks the passages of the lungs and the pancreas."

The discovery of the CF gene had several implications of almost immediate importance to medical science. Once the team had found the other error or errors in the same gene that explained the remaining 30 percent of CF cases that did not show the mutation they had already found, new genetic tests could be created that would determine conclusively whether or not any individual is a carrier of the gene. Even more exciting, the discovery of the gene could be expected to lead quickly to the isolation of the faulty protein, and ultimately to new treatments aimed at the cause of cystic fibrosis, not merely at its consequences.

In 1985, as CF researchers were beginning to make a map of those 2 million nucleotides surrounding the CF gene, a few biologists were talking openly about an idea that made the efforts of Kunkel and the Collins-Tsui team sound almost mundane. It now seemed feasible to make a map of the entire human genome, of all of the DNA in all 23 pairs of chromosomes, and then to deduce its complete sequence of about 3 billion nucleotides.

Five years ago the idea sounded crazy, or visionary. Today, it is reality. By fiscal year 1990, the U.S. government was devoting about $87 million to such a project, to be divided between several government agencies. Some 1,500 hereditary conditions were pinned to the human gene map.

The term *human gene map* was, by then, more than an abstraction. A year before that, a preliminary map was in print, in the scientific journal *Cell*. It was a list of locations of more than four hundred genetic markers, landmarks on the human chromosomes so closely spaced and well distributed that any geneticist had 95 percent odds of finding the correct general location of a human gene, as Susan Chamberlain and Lap-Chee Tsui had done.

At first, the "human genome project" was only an informal network of collaborators. Collectively, they and thousands of other geneticists are advancing toward a much broader goal than the mere solution of a few thousand rare hereditary diseases. Within a few decades, almost certainly, we will know the locations of most of the important genes in the human genome.

Today, there is a National Center for Human Genome Research within the U.S. government. There is a set of codified national research programs in several other countries, aimed at mapping parts of the human genome. There's even an international organization, the Human Genome Organization, which is sort of a United Nations of human genetics assigned to coordinate the various national projects. This is not science fiction. It's the future.

The research is moving so swiftly now that any attempt to chronicle the field could not be up to date. New genetic linkages are reported by the week. Genes will have been discovered by the time these words are in print that were unknown when they were written.

The creation of the program was spurred by two favorable reports in 1988. One came from the National Academy of Sciences, and strongly supported the creation of a so-called human genome project. "An improved capacity to identify genes related to disease will have an immediate impact on the diagnosis, treatment, and prevention of genetic disorders," wrote the authors.

Another report, in the same year, came from the U.S. Office of Technology Assessment, an advisory arm of the U.S. Congress. "At current or perhaps increased levels of effort," it said, a genome project "may eventually make possible control of many human diseases—first through more effective methods of detecting disease, then, in some cases, through development of effective therapies based on improved understanding of disease mechanisms."

A more complete understanding of the human genetic message, both reports pointed out, should also reveal a great deal about how the healthy human body works. The genome project gives scientists new access into cloistered regions of the body such as the brain, because it should lead to the discovery of many yet-unknown body chemicals, as it pointed to dystrophin and the CF protein. Besides explaining the causes of disorders, this will reveal a great deal about how we think and act.

The estimated total cost of the organized effort to determine the contents of our DNA by around the turn of the century is about $3 billion. This sounds vast to most individual biologists. But it is slightly less than the federal government spends on computers in a single year. It would buy only two or three nuclear submarines or aircraft carriers. For that price, we can possess the original owner's manual to the human body, now written in terms we can comprehend.

The American Cancer Society estimates that altogether we spend more than twenty-three times as much in a single year—about $70 billion—caring for people who have cancer. Genome research has already begun to reveal ways to predict cancer genetically. Eventually, we can hope, doctors will be able to use genetic tests to trace and eradicate cancer, or even to prevent it before it requires any care whatever.

A century ago, doctors bound wounds and wiped foreheads, but often they could do little else for a patient except sit and wait. Today we fall ill, and then we expect to take a medicine and recover. For the ailments that still defeat many of us—cancer and heart disease, in particular—doctors still respond mostly with crisis management and damage control.

In the near future, they may be able to tell us what's coming before it strikes, by looking at our genes. As a result of the genome project, its champions predict, the medicine of the next century may be much less reactive, much more preventive.

This may be good news, or bad news, or both. As the British physician and gadfly columnist Donald Gould put it in *New Scientist*, the genome project may be seen as "a device for letting the 'healers' get their hooks into you a bit sooner, and for increasing the profits of the pharmaceuticals industry and for generally boosting the medical trade." Later he added: "I don't *want* to know what's going to happen. It'd scare me stiff."

At best, however, medical research may eventually provide enough information, based on studies of human genes, to help anyone avoid or minimize the disease dealt him by his genetic legacy. Meanwhile in the indefinite period between the discovery of a genetic linkage and the clinical progress that follows, we will be the ones, like Darren LeBlanc and Zoe Weaver, who sit and wait.

Then many more of us will know firsthand Zoe's mixture of optimism and caution in the face of a known risk. When people ask what it was like to watch the discovery of the CF gene, she says, "I always feel bad, because the way people ask, it seems like they want me to say we were all jumping up and down and breaking out the champagne. That's not the way it works.

"I feel very hopeful," she goes on, "but because I've been working in the lab, I realize how much more there is to do. . . . Just because you have the gene doesn't mean that much. It's how you apply that knowledge, and it always takes a while to know how to use it.

"First you have to be able to characterize the protein; then there's, How does the protein work? And then there's the whole gene therapy area, where there's a whole lot of work left to be done. Nobody really knows exactly how to insert the correct gene into someone."

Again, Zoe cannot end her words with the present reality. She changes course: "But I think there's really a lot of promise. I mean, the research has been going at a very fast pace, and even though we don't have a cure today, a major blocking point has been gotten over."

The project to map the human genes is sometimes called a revolution in science. It is not. The discovery of the molecule behind heredity and the recognition of its structure and operation, in the 1950s and 1960s, were the scientific revolution. The genome project is only its technological outcome.

However, an intimate knowledge of all our genes can be expected to cause developments in society that are revolutionary, at once threatening and promising. The rise of computers in the last decade is a perfect parallel. Everyone can see their pervasive and momentous impact on daily life. The results are sometimes useful and even fascinating—but also occasionally irritating, potentially very dangerous, and already taken for granted.

3

HERRICK'S PUZZLE

The first human disease to be understood in molecular detail was a disorder of that easily obtained fluid, the blood. The search for its cause was a trek backward down the genetic pathway. It proceeded from a very sick person to his obviously abnormal cells, from there to a seriously faulty protein, and at last to an error in a gene—a very simple error.

At the outset, we have only the records to go on. In September 1904, a black man of twenty disembarked from a ship in New York City, en route from Grenada in the West Indies to Chicago. He had an open ulcer the size of a half-dollar near one ankle, for which he sought medical attention. It was treated with iodine and healed, leaving a scar—the latest of many on his legs. The man proceeded to Chicago.

Sometime in November he developed a persistent cough, and on Christmas Eve the cough suddenly worsened and he felt feverish. On the day after Christmas, with a fever of 101 degrees Fahrenheit, he went to a hospital. There he was seen by the attending physician, Dr. James Herrick.

Herrick was a tall, patrician man of forty-three, a Republican and a Congregationalist, a professor of medicine whose favorite diversions were gardening, golf, and reading Chaucer. He had a keen interest in medical research, and was taking chemistry courses at the University of Chicago. To judge from his photograph, he would have spoken gently to his patient.

Of the patient, we have Herrick's own description: An "intelligent Negro of twenty" with a "typical Negro" face and black, curly hair. "He was fairly well developed physically," Herrick added, "and was bright and intelligent," having come to Chicago for postgraduate education.

Herrick began with a physical examination. He looked at the patient's gums, which were pale, and his eyes, which were jaundiced. He tested his hearing, examined his ears and his nose, which were inflamed, and felt enlarged lymph glands. Then he noticed the scars. He twisted the man's legs and lifted them; there were scars on the calves and thighs as well.

"There were perhaps twenty scars in all," he wrote later. "They were rounded or oval, sometimes of irregular contour. Some were like tissue paper or thick parchment to the touch and were lighter in color than the surrounding skin. They were strikingly like scars often seen as the result of syphilis."

"How do you think you came by these scars?" Dr. Herrick asked the young man.

The patient replied that he always assumed they had come from running bare-legged through the brush.

"Did you ever have venereal disease?"

"No, sir."

"Ever feel pain or itching in your private parts?"

"No, sir."

Herrick grunted, and proceeded to lift his stethoscope to the young man's chest. The heart thundered into his ears.

"Have you been taking any medication?"

"No, sir."

Dr. Herrick listened again.

"When did you have your last drink?"

"You mean liquor, sir?"

"Liquor, yes."

"I don't take liquor, sir."

Doubtless Herrick formed his opinion of the man's intelligence as he listened to him tell his own history. His father had died in an accident; his mother and three siblings were still living in Grenada. He reported only one important childhood disease: yaws, a parasitic tropical disease characterized by skin lesions, which he contracted at the age of ten and which took nearly a year to heal.

Since the age of seventeen, the patient told Herrick, he had noticed palpitations and shortness of breath and what the doctor recorded as a "disinclination to take exercise."

Many of the symptoms—the cough, the anemia and jaundice, the leg ulcers—pointed to an infection of some kind. The young man's Caribbean origins pointed to an unusual one. Herrick asked the man to provide samples of his urine, stool, and sputum, all of which were tested several times during the next few weeks, while the patient rested in the hospital. Herrick never saw any sign of infection.

In fact, the only unusual detail he could find was something so strange Herrick doubted it was real, and so he looked for it again and again. When he examined the patient's blood under the microscope, to document the anemia, the shape of the red cells was quite remarkable: "very irregular," he wrote later, "but what especially attracted attention was the large number of thin, elongated, sickle-shaped and crescent-shaped forms."

At first, Herrick assumed the strange cells meant something was wrong in the laboratory. He took blood from several other people, and it was all normal. But nothing done to the patient's blood—not heating it, diluting it with alcohol or ether, or staining it with dyes—made the red cells revert to normal appearance.

"The elongated forms as well as those of ordinary form seemed to be unusually pliable and flexible," Herrick noticed, "bending and twisting in a remarkable manner as they bumped against each other or crowded through a narrow space and seeming almost rubberlike in their elastic resumption of the former shape. One received the impression that the flattened red discs might by reason of unusual pliability be rolled up as it were into a long, narrow bundle."

The patient left the hospital after four weeks of rest, nourishment, and iron supplements for his anemia. At the time of his release, his blood corpuscles still assumed the peculiar shape.

Two years later, again on the day after Christmas, the young black man was readmitted to a hospital, this time with what Herrick called "muscular rheumatism": pain in the back, legs, and arms, as well as pallor and slight fever. He was in the hospital for two months, and departed apparently healthy. That was the last Herrick ever knew of him.

The red-cell deformation Herrick had seen was not entirely new. It already had a peculiar-sounding medical name: poikilocytosis. It had been reported in 1840, by the assistant surgeon to the British Royal Regiment of Horse Guards, in several species of deer. But Herrick considered its presence in human blood so remarkable as to warrant a formal presentation at a medical conference in 1910, and a subsequent detailed report in the *Archives of Internal Medicine*.

Although it became known as Herrick's anemia, the man who discovered it was at a loss to account for it. "No conclusions can be drawn from this case," he wrote. "Not even a definite diagnosis can be made. Syphilis is suggested by many of the facts. . . . The scars said to have been due to yaws were like those left by syphilis. . . . The question of diagnosis must remain an open one unless reports of other similar cases with the same peculiar blood shall clear up this picture."

Herrick proceeded to an illustrious career: he was one of the first to use an electrocardiograph machine, and was the first to describe coronary thrombosis in a human being. The anemia he had discovered eventually went under another name, sickle-cell disease, and its discovery barely rated a mention in Herrick's own memoirs. Nearly twenty years after the discovery, no more than four additional cases had been documented.

In West Africa, the ailment was well known among ordinary people, who called it by still other names: *chwech-weechwe, nwii-wii, nuidudui,* gnawing sounds that apparently evoke the way the illness feels, near the bones. Literally translated, *nuidudui* means "body chewing," which is the way African patients often describe the sensations of a sickle-cell crisis. An American will say he is on a rack.

It is now known that sickled red blood cells cannot circulate freely in the small blood vessels supplying the muscles, and therefore fail to supply them with enough oxygen. This leads to

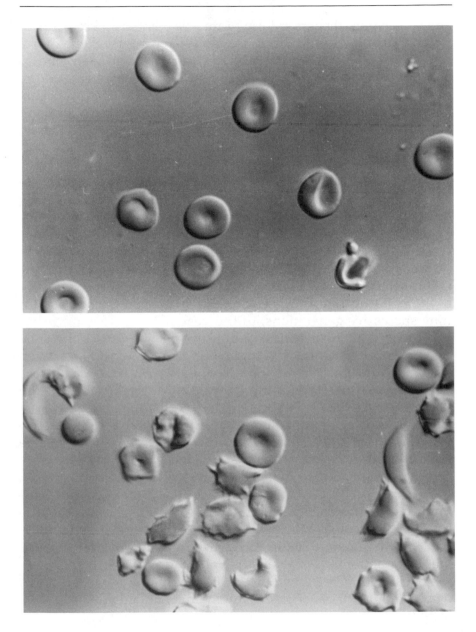

Normal red blood cells (top) and sickle cells.

an excruciating, dispersed form of muscle cramps—a kind of pain so unique and exquisite that it is no longer any challenge for a doctor to identify it. In the sickled form, red cells also tend to accumulate and clog small blood vessels, which can lead to the destruction of almost any organ in the body. The anemia occurs largely because the strange shape of the sickled cells makes them very prone to mechanical destruction.

"Our forefathers . . . knew not only that the disease 'ran through families,' but also that in the vast majority of cases parents of chwech-weechwe children were not [visibly] different . . . from normal adults," wrote a Ghanaian doctor, Felix Konotey-Ahulu, in the *Archives of Internal Medicine* in 1974. "More remarkable was the ability of our forefathers in Ghana to recognize that when one of the parents was a sufferer the proportion of children who suffered from the disease was . . . greater than usual." Konotey-Ahulu reports that in one family of a Ghanaian tribe, sickle-cell anemia has been traced back for nine successive generations, to the year 1670.

In some parts of Africa, one in five people carries the gene for sickle-cell anemia. The figure among American blacks is lower: about eight in every hundred. Only two American blacks in a thousand has two of the genes, and therefore the disease itself. Nonetheless, as reports of sickle-cell anemia began to accumulate slowly in the United States (following a 1922 description in the *Journal of the American Medical Association*) the erroneous impression arose that sickle-cell anemia was more common in the United States than in Africa.

As late as 1950, one "expert" declared that "some factor imported by marriage with white persons is especially liable to bring out the . . . disease, while the anomaly remains a harmless one in the communities in which it originated." He used this fallacy as an argument against racial intermarriage.

He could have known better. During the previous year, the true hereditary basis of sickle-cell anemia had already been published.

James V. G. Neel earned his Ph.D. in genetics in 1939, and then in 1944, to amplify his understanding of medical genetics, became a doctor. Neel encountered his first case of sickle-cell anemia as a staff doctor in Rochester, New York. "It was a problem screaming to be looked at," he says, "a mysterious and ill-defined disease." It was quite clear by then that the disease ran in families, but not at all clear how it was inherited.

As Neel entered genetics, the field was falling under a shadow. America was at war against a regime that represented the apex—or, more aptly, the nadir—of eugenics, the effort to direct human breeding in order to try to improve the "quality" of human populations by eradicating certain traits. The Nazis were masters at it. Before they decided to exterminate the Jews, the Nazis adopted a law that forced sterilization on certain "undesirables," including people with schizophrenia, blindness, and epilepsy. Later, they began to exterminate institutionalized people so that German society would not have to pay for "worthless lives."

When Neel came into medical genetics, he says, he was determined "to avoid the opprobrium and taint of work done in the name of eugenics." He intended to stay out of the quicksand beneath the favorite concepts of eugenics. Its proponents regularly spoke about the heredity of value-laden qualities such as morality, resourcefulness, and productivity as if their inheritance were established fact.

Because of these dangerous habits, Neel resolved to use the strongest, most specific scientific methods available in his own research, to define clearly what he was talking about, and to decide carefully beforehand what he would be looking for. These were precisely the strategies needed to clear up the confusion about sickle-cell anemia.

Researchers had learned that red blood cells taken from patients with sickle-cell anemia would assume the crescent shape whenever the cells were deprived of oxygen. Some dark-skinned people—often the parents of sickle-cell anemia patients—appeared to have "latent" disease. Their red cells would also sickle when deprived of oxygen, but these people appeared to be healthy. It was unclear whether people who had some sickled cells but appeared healthy merely had a mild form of the disease. It was also in dispute whether the disease was dominant or recessive, or something entirely more complicated.

There were too few blacks in Rochester at the time to allow Neel to start a family study. But when he moved to the University of Michigan in 1949, he almost immediately began to look

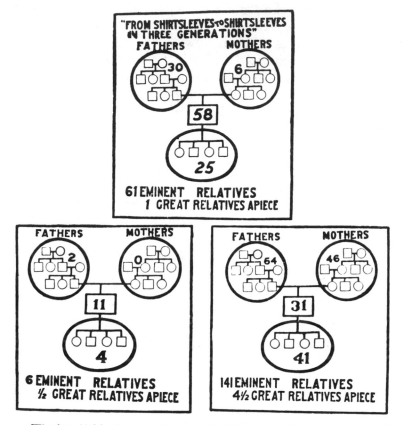

"FROM SHIRTSLEEVES TO SHIRTSLEEVES
IN THREE GENERATIONS"
FATHERS   MOTHERS

58

25

61 EMINENT RELATIVES
1 GREAT RELATIVES APIECE

FATHERS   MOTHERS

11

4

6 EMINENT RELATIVES
½ GREAT RELATIVES APIECE

FATHERS   MOTHERS

31

41

141 EMINENT RELATIVES
4½ GREAT RELATIVES APIECE

Whether "shirtsleeves return to shirtsleeves in three generations," or not is seen to depend on whether or not the good blood is kept in the family. The upper left-hand circles show the number of distinguished ancestors on the side of the fathers, the upper right-hand circles, the number on the side of the mothers, and the lower circle the number of distinguished descendants left by the distinguished men noted in the central square. It will be seen that the number of distinguished descendants decreases in proportion to the amount of distinction among the ancestors.

Pedigrees of distinction. Like many eugenicists, the author failed to define "shirt-sleeves," "eminent," "great," and "distinguished," while implying that they are purely genetic in origin. Reprinted with permission of Macmillan Publishing Company from *The Fruit of the Family Tree* by Albert Edward Wiggam, copyright 1926 by The Bobbs-Merrill Company, Inc.; copyright renewed 1954 by Albert Edward Wiggam.

for kindreds with sickle-cell anemia, both in the University Hospital in Ann Arbor and through three Detroit-area hospitals.

To test the theory that sickle-cell anemia might be a predisposition that flared up and disappeared, it was necessary to trace patients and their families to their homes, and study their blood while they were feeling well. Neel contacted families with the help of a black social worker. His success was spotty. "Family studies are never easy, and this was a little more difficult than most," he says. He was starting with hospital records that were ten years old. Many families had moved in the interval. Broken homes also made it difficult to trace some individuals. When contacted, people often could not see the need for any special studies of a child who appeared to be well.

Nonetheless, Neel managed to examine 465 individuals in a total of 75 families—far more patients at once than anyone else had studied. The real keys to Neel's success were the facts that he was the first trained geneticist to address the question of the heredity of sickle-cell anemia, and that he used four different tests to establish whether red cells would sickle. He stimulated sickling methodically, exposing the cells to chemicals that would rob them of oxygen, and then watched them for one or two days, not just a matter of hours.

In a preliminary report in 1947, Neel revealed that in every one of 29 cases of sickle-cell anemia, both healthy parents had some red blood cells that would sickle under the proper conditions. Considering the patterns of heredity in his kindreds, Neel reported, "the most reasonable hypothesis which will render the above data intelligible is that the sickling phenomenon is due to a gene which in single dose (heterozygous condition) produces only the sickle-cell trait, and in double dose (homozygous condition) sickle-cell disease."

As some African tribes had known for centuries (according to Konotey-Ahulu), if two heterozygotes for the sickle-cell gene— two healthy carriers, with a single copy each—conceived a child, the fetus would have one chance in two of being a carrier itself and one chance in four of having the disease. If a homozygote, someone with the disease, married a carrier with only one gene for sickle cell, the odds were 50–50 that any child of the union would also have sickle-cell anemia.

Neel published an expanded version of his sickle-cell study

in 1949. Only six months later, his explanation was confirmed and there were the beginnings of a molecular explanation for the "mysterious and ill-defined disease."

Had James Neel wanted to search further for the gene behind sickle-cell anemia, there would have been nowhere to go. The field of genetics was still too young. On the other hand, of the protein hemoglobin, which carries oxygen inside red blood cells, there was intense study. One of its most serious investigators was the biochemist Linus Pauling, of the California Institute of Technology.

Pauling became interested in the sickle-cell problem because of a chance conversation with William Castle, a specialist in blood diseases from Harvard University, during a meeting of a U.S. presidential commission on medical research in 1945. Castle described to Pauling the way that red blood cells changed shape when they were deprived of oxygen. In an instant, an explanation sprang into Pauling's mind.

"I pointed out that the relation of sickling to the presence of oxygen clearly indicated that the hemoglobin molecules in the red cell are involved in the phenomenon of sickling," Pauling recalled later, which ". . . could be explained by postulating that [sickle cells] contain an abnormal kind of hemoglobin, which when deoxygenated has the power of combining with itself into long rigid rods, which then twist the red cell out of shape."

Pauling had been busy studying how molecules in the immune system fit together like locks and keys, so he was primed to think of molecules as Tinker Toys. Proteins in the blood are designed to remain soluble, he knew, not to stick together. "It was accordingly easy to have the idea that the hemoglobin molecule might be changed in such a way as to produce two complementary regions," he recalls. They might clump together, and deform the red cell.

In the following year, Pauling wrote to Harvey Itano, an incoming graduate student, that he was interested in having someone in the lab study the chemical nature of hemoglobin. Beginning in 1946, Itano analyzed numerous properties of sickle hemoglobin, most importantly the way the molecule traveled in an electrical current.

The chemistry beautifully supported Neel's genetic hypothesis. Itano's studies ultimately showed that hemoglobin from a normal person traveled in one direction in a solution through which an electrical current was passed. Sickle hemoglobin traveled the opposite way; it was chemically different. Hemoglobin from healthy parents of sickle-cell patients traveled both ways. Evidently they possessed both normal and sickle hemoglobin.

In later years, Neel went on to discover many new and unusual forms of hemoglobin all over the world, and Itano proceeded to differentiate them chemically. At first they were identified by letters of the alphabet. Hemoglobin S was the form that sickled. Hemoglobin C, discovered in Detroit, was common in West Africa. Hemoglobin D, first isolated in Los Angeles, is prevalent in India and hemoglobin E in southeast Asia. The letters of the alphabet ran out quickly. By now, more than four hundred variants—polymorphisms of the hemoglobin gene—are known to exist. Most are harmless; a few of them cause diseases known collectively as the thalassemias (after *thalassa,* the Greek word for "sea," because many patients come from Mediterranean cultures).

The defect was clear from the start: Hemoglobin S sickled. Neel, Pauling, and Itano clarified the genetics. It was two British researchers, Vernon Ingram and Max Perutz, who carried out the next step: discovering what was faulty about the protein.

Ingram was intrigued by the idea of verifying a theory central to the new genetics: that a change in a single DNA nucleotide could cause an alteration in a protein. He chose sickle hemoglobin as a way to prove the point. Ingram used chemicals to break the long hemoglobin molecule into smaller pieces. Then he forced these pieces to spread out across sheets of filter paper by exposing them to different solvents and different electrical fields. This had the effect of separating them both by size and by electrical charge to create a scatter pattern.

The snippets of sickle hemoglobin made a subtly different pattern from those of the normal protein. Eventually, Ingram was able to prove that they differ by as little as a single protein subunit. Where normal hemoglobin has a glutamic acid molecule, sickle hemoglobin has valine. It is that simple. The result is disaster.

In an adjacent laboratory in Cambridge, England, Max Perutz was learning the reason why. He had been subjecting hemoglobin to intensive study with X rays, using mathematical analysis of reflection patterns to deduce the three-dimensional structure of the protein. During these studies (for which he and chemist John Kendrew won a Nobel Prize in 1962) Perutz showed that the aberrant amino acid sat on the surface of sickle hemoglobin, where its abnormal electrical properties were critical.

The change made the molecule electrically sticky; now it would insert itself into a pocket in another hemoglobin molecule, a cavity into which the normal glutamic acid would not fit. Hundreds of other single mutations alter the hemoglobin molecule harmlessly, but at this point on the surface of hemoglobin there is no room to spare. Inside the red blood cell, hemoglobin molecules are separated from each other by barely more than the distance between atoms in molten metal. A minuscule alteration—a single amino-acid variation near the surface—leads them to form a gel and, ultimately, a crystal.

By the 1970s, electron microscopes had revealed to the human eye the truth of what Linus Pauling had predicted: In the absence of oxygen, sickle hemoglobin polymerizes to form a double stranded fiber, which in turn adheres to other fibers as part of a 14- or 16-stranded twisted rope. As Pauling would have it, a rod. Or even, as James Herrick observed in 1904, a long, narrow bundle. The red blood cell deforms, and grows rigid.

Vernon Ingram used sickle hemoglobin to prove a point: that a variation in a single amino acid, and by inference a very minor error in a gene, could lead to the deformation of a protein. It took many years for that information to be converted into something of practical use to patients. But ultimately, sickle-cell anemia became the first disease to be detected by molecular variations in the DNA, rather than by grossly visible aberrations of entire chromosomes.

The person who pioneered this test was Yuet-Wai Kan, a hematologist who emigrated from Hong Kong to Canada in 1960. In Montreal he encountered a patient with thalassemia, and began to study the disease. At the time, the clinical varieties of the thalassemias were becoming clear, but the molecular explanation—the many varieties of mutant hemoglobin—had not yet been discovered.

In 1972, not long after he joined the University of California at San Francisco, Kan learned from an obstetrician about the new technique of fetal blood sampling. In a few medical centers, specialist obstetricians had been using a new instrument—a combination of microscopes and optical fibers—to obtain blood samples directly from the scalp of a fetus inside its mother's uterus. Kan began to explore the possibility of obtaining fetal blood as a way to diagnose the thalassemias before birth.

"At that time," he says, "the biggest problem was obstetrical. It was not that easy to get fetal blood. The sample was not usually pure, and you had to purify it from the maternal blood." Besides, the method could hardly become widespread; it required very special expertise and sophisticated equipment.

Instead of focusing on the protein, hemoglobin, in fetal blood, Kan gradually shifted his attention to the corresponding genetic material, the DNA, in adult blood cells. (Researchers in Boston had isolated the hemoglobin gene in 1978, after using the genetic code to deduce the DNA sequence that corresponds to the amino-acid sequence of hemoglobin.)

Kan was interested in studying the linear structure of the hemoglobin gene. He began trying to create overlapping fragments that would allow him to piece the structure together, by deduction, something like the way you might fit together the pieces of a torn and discarded letter.

Working with molecular hematologist Andrée Dozy, Kan began by tearing the invisible DNA into pieces, using an *enzyme*—a biological catalyst. Biologists have been using enzymes extracted from bacteria for generations (beginning with the isolation of the ones that ferment beer). Kan and Dozy employed quite a new variety, called *restriction enzymes*. These enzymes also come from bacteria, which use them as a defense against viruses.

In nature, restriction enzymes exist to chop a viral DNA chain into bits. But they don't just blast the DNA into random fragments; they nip it only in specific places, between particular pairs of nucleotides. Thus they chop long DNA messages into pieces of varying but predictable size. Biologists use them for scissors.

It's as if an error in the printing process inserted a space in the middle of every sequence *in* on this page. The previous sentence would look as follows:

> It's as if an error i   n the pri
> nting process i   nserted a space i
> n the middle of every sequence "i
> n" on this page.

If some copies of this book had a typographical error deleting the *n* from the word *inserted,* the error could be found by chopping between every "in" sequence on this page in all the books and comparing the fragments. It's a most tedious way to look for a typo, but it was the best trick biologists had at the time for detecting errors in DNA. (They still don't know how to read the genome like a book; they have to deduce its message letter by letter.)

Small differences in the DNA of different individuals lead to obvious variations, or polymorphisms, in the way restriction enzymes break human DNA into fragments. Kan and Dozy discovered one fragment that appeared when enzymes broke up DNA from normal people, but not when they used them on DNA from people with sickle-cell anemia or from carriers.

In their first published attempt at using this for diagnosis, Kan and Dozy used the difference between DNA fragment sizes to determine that a fetus conceived by a couple in Dallas did not have sickle-cell anemia, but was a carrier. The mother carried the fetus to term, and delivered a healthy baby.

"Then we realized how useful this method could be," Kan recalls. "What we did was we studied several families with sickle-cell anemia. We found the marker [the difference in fragments] was consistent."

Not perfectly consistent, however: occasionally Kan failed to find the expected fragment pattern in DNA from someone who had the disease. It was obvious that the enzyme was not breaking the DNA precisely where the sickle-cell mutation occurred. It was breaking the DNA in a marker located somewhere near the gene. Sometimes, albeit rarely, the marker and the mutation were separated by accidents in the process by which the chromosomes duplicate themselves, and a misleading pattern arose during DNA testing.

The main problem with Kan and Dozy's initial test was that the "typographical error" that the enzyme detected in some DNA with the sickle-cell gene was not the same error that caused the disease itself. But because they knew the amino-acid sequence of the hemoglobin protein by that time, and could deduce the sequence of the sickle-cell gene and its normal counterpart using the genetic code, Kan and Dozy and anyone else who was interested could identify exactly the subunit sequence they needed in order to detect the gene itself using a restriction enzyme to disrupt: CTGAG, altered to CTGTG in the mutant gene.

A few years later a suitable restriction enzyme was isolated. In 1982 Kan and a co-worker, Judy Chang, and independently

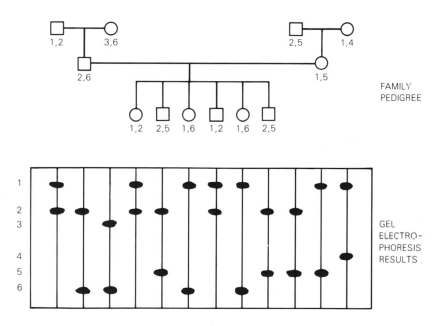

FAMILY PEDIGREE

GEL ELECTRO- PHORESIS RESULTS

Tracking the inheritance of a restriction-fragment-length polymorphism. Depending on which restriction-enzyme cutting sites people have inherited, their DNA will break into fragments of different sizes, which can be measured using gel electrophoresis. The fragment sizes are inherited as predictably as traits such as eye color or right-handedness. (Adapted from Ray White and Jean-Marc Lalouel, "Chromosome Mapping with DNA Markers," *Scientific American*, February 1988) two other groups, reported developing a still better test, using an enzyme that sliced the gene in the exact location of the error that causes sickle-cell anemia.

From that point on, a genetic test for sickle-cell anemia could be absolutely accurate, as long as people performed it correctly. Not only is sickle-cell anemia understood at its most fundamental level, it is now detectable at that level too.

Today prenatal tests for sickle-cell anemia and the thalassemias are routine, are based on detection of the faulty gene, and are available anywhere in the United States. Sickle-cell anemia is one of about seventy-five hereditary diseases that can be detected before birth by genetic testing.

In recent years, restriction enzymes have provided many more new prenatal tests for (among others) the degenerative nerve disorder Huntington's disease, muscular dystrophy, and cystic fibrosis. As with sickle cell a decade ago, the prenatal detection of cystic fibrosis is shifting rapidly from genetic marker

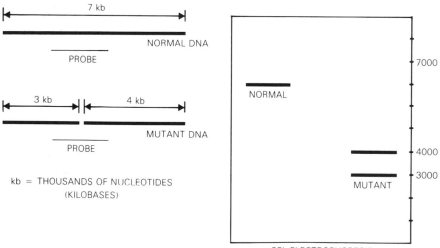

Diagnosis by DNA. The mutant DNA creates two small fragments, the normal DNA only one larger one. The difference is detectable with electrophoresis.

analysis (almost perfect) to detection of the faulty gene itself (perfect). Month by month, the list will continue to grow, as the genes behind more diseases are hunted down and put on the map.

But the implications of gene mapping and DNA analysis reach far beyond prenatal testing. More and more, when something is wrong with someone we will know why—right down to the molecules. We will also learn more and more, steadily, about what goes on when nothing is wrong.

4

THE CASE OF THE HYPOTHETICAL CULPRIT

There's far more to genetic linkage studies than white coats, test tubes, and molecules. These studies begin, as they must, with ordinary people. Seeking out their human subjects may take researchers to unfamiliar places where they may find themselves pleading for blood from total strangers.

Outside, a sunny autumn day was waning, one of those crisp days that awakens you to the pleasures of the season and reminds you that winter too can be bright. Inside, there was a vaguely distasteful odor, human and stale. A man continued to pass his remaining days, asleep. He was flat on his back on a well-made bed, fully dressed except for shoes. Let's give him a name, a Dutch-sounding name. Henry van Dyke.

An unexpected disturbance: a man approached. We'll refer to him as William Holbein. (William and Henry might not mind revealing their real names, but other people in their extended family might resent being identified with such a bizarre subject.)

"Henry? Are you awake, Henry? There are some people

76

here to see you.'' Henry stirred. William shook him gently. ''Wake up, Henry. These doctors would like to visit you for a while.''

Dr. Roger Kurlan slipped behind the side of the bed near the wall, and bent down. ''Are ya awake, Mr. van Dyke? Could ya wake up? We've come all the way ta Michigan ta study yer family, and we'd like ta talk to ya for a few minutes.'' (This was his bedside manner of speech, broader and more colloquial than the way he talks to friends, or to other doctors.) ''Could ya wake up?'' Henry's eyes fluttered. ''Can ya open your eyes?''

One eye opened slowly, luminous, inside a puffy lid, a lizard's eye.

''We're here to study a medical problem in your family that causes tics and twitches and eye blinks, Mr. van Dyke. Can ya hear me?'' Henry nodded. ''Can ya sit up?''

''Hurts ta sit up,'' he rasped.

''Where does it hurt?'' Kurlan responded, the way a son would ask.

''M' legs,'' Henry slurred, and tapped a thigh with his palm.

''Do the soles of your feet hurt?''

''Not the soles. M' legs. Down to the bottom.''

''Down to the bottom,'' intoned Peter Como, who had come along the other side of the bed.

''Are ya sleepy, Mr. van Dyke?'' asked Kurlan. ''Can ya open your other eye?''

''Always sleepy,'' he replied.

''Always sleepy,'' noted Kurlan. ''Are ya on any medication?''

Henry nodded. ''Fifteen pills.''

''Fifteen pills a day, Mr. van Dyke?'' Kurlan raised his eyebrows and looked to the foot of the bed, toward Laurie Ernest, the nurse with the research team. He pointed out the door. She nodded, did a right face, and left.

''Mr. van Dyke, as I said, we're studying this medical problem in your family, where people have tics and twitches and blink their eyes all the time.'' Kurlan called out the words as if he were shouting over a chasm. ''Do you remember anybody in your family, Mr. van Dyke, any of your brothers and sisters who did that?''

Immediately Henry nodded. ''Emma,'' he whispered. ''Used ta blink her eyes.''

''Emma. That's exactly right,'' said Kurlan. ''Anybody else, Mr. van Dyke?'' A shadow wafted over Henry's face; Kurlan waited. ''How about your parents, Mr. van Dyke. Your mother or father?''

''Mother used to blink her eyes. And twist her shoulders.''

''That's right, Mr. van Dyke,'' Kurlan said. ''What's this in your pocket? Is this your glasses?''

''M' teeth.''

''Can you put them in, Mr. van Dyke?'' Henry merely opened his mouth and waited; Peter Como leaned forward, cradled Henry's head, and wrestled the dentures in. Then he lifted Uncle Henry to a sitting position, feet on the floor. Henry slumped forward.

Kurlan moved out from behind the bed and came to kneel in front of him. ''How are ya, Mr. van Dyke?'' he asked. ''Are ya pretty awake now? How d'ya feel?''

Henry stared down, motionless. Laurie Ernest slipped back into the room and handed Como a note: the medication. ''Amazing,'' she whispered. ''They just handed me his chart.''

Kurlan was gazing at Uncle Henry. ''Well, we won't need to bother you for too long. But we hope you can give us a little help. What we're tryna do is see as many of your relatives as we can. See who has twitches and who doesn't. The reason we came here is, you have such a big family, so many so close together. We want to get as many blood samples as we can take back to New York, to see if we can find out what's causing these twitches.''

As Kurlan spoke, Henry was twisting to look over his shoulder. ''Who's here?''

''Why, that's Laurie, our nurse practitioner. And Dr. Como, our neuropsychologist, and me. That's all.''

''Where's William?''

''He's gone to wait in the lobby. Why? Shall I get him?''

''No. I didn't want to say while he was here. He's got it too.'' Henry shrugged and twisted his neck, imitating the way William shifts inside his collar, again and again, habitually.

''Where'd ya say you're from?'' he asked, and at last looked directly at Kurlan. With both eyes open and his teeth giving shape to his face, he was handsome: Crystalline eyes, strong cheekbones, silver hair.

"We're from the University of Rochester in New York. Actually, I have something I could show you about us. Do you know where your glasses are?"

Mr. van Dyke shook his head. So Kurlan recited to him the relevant information.

Henry van Dyke belongs to one of the world's largest known kindreds affected with Tourette's syndrome. This was the first time Henry van Dyke had heard the name Tourette's, but he knew the syndrome well. Over the years he had noticed and studiously ignored some socially embarrassing traits in some of his relatives: repetitive shrugs and twitches, convulsive blinks, intrusive whoops and belches.

When he was young, fifty years ago, Henry and everyone else would have called these things personal peculiarities, odd habits, mannerisms. Children might point out the mannerisms and laugh at them; adults would have let them pass unremarked, taken them for granted. They were just another small part of the collective personality of a midwestern community—no more and probably less remarkable than Christian virtue, facial hair, or musical talent.

For about the last decade, however, medical scientists have been paying new attention to the manifestations of Tourette's syndrome, a century after the French scientist Georges Gilles de la Tourette first categorized them as an entity. In recent years, Tourette's syndrome has been taken out of the closet and dusted off, redefined as a "condition," dignified as something worthy of recognition and even study.

It's clearly hereditary. Kurlan and his collaborators are banking on the likelihood (to judge from the familial pattern) that a single gene causes Tourette's, most likely a dominant gene. If so, whoever inherits a single copy of the hypothetical gene, from either parent, will probably develop some signs of Tourette's. The research team is in Holland, Michigan, for the second time, still on the trail of this gene.

"An understanding of the pattern of inheritance . . . may provide clues to better methods of treatment," Kurlan's informed-consent form says. ". . . Although heredity appears to be an important factor, a specific pattern of transmission has not been determined at this time. For these reasons, it is important to study large family groups where there already exist some identified cases of Tourette's syndrome."

"Henry" and his relatives are particularly useful because (for reasons none of them seems able to explain) they have nearly all remained within about a fifteen-mile radius of Holland, the town their Dutch forebears founded near the shores of Lake Michigan more than a century ago. Given that the syndrome is probably dominant—a single copy of the hypothetical gene may cause the syndrome—inbreeding would not be required to explain its prevalence in the family, as it would be if a recessive disease were unusually prevalent. What mattered to Kurlan and his team was the family's concentration in a small area. All it took to study the family in detail was a quick trip to the Grand Rapids airport, a rented station wagon, and a few long days of work.

They were hoping to link the symptoms to some genetic marker. The validity of such genetic linkage studies increases dramatically with every new blood sample from a close relative that can be analyzed. In addition to what they had already gathered in Michigan and elsewhere, the researchers hoped this return visit to Holland would provide enough blood samples to document the location of a gene that causes Tourette's syndrome. If such a gene exists. If the competing researchers studying Tourette's can resolve which traits are part of the syndrome (among a list of strange behaviors that seems to be expanding almost indefinitely) and which are incidental or irrelevant.

As Kurlan read the consent form aloud, Henry began to lean to the left. Kurlan leaped to return him upright, then sat on the opposite bed, where he could look straight across at him.

"How about yourself? Did you ever have any of those tics or twitches yourself?"

"No."

"How about your speech? Did you repeat things? Did you stammer?"

Henry nodded.

"What was it like?"

"D-d-d-d," he said. "S-s-s-s."

"Did you ever have any words you liked to say to yourself, or out loud? Any swear words?"

Henry sat motionless for a moment, then shook his head.

Standing next to Kurlan, Peter Como watched with a look of gentle gravity. "Mr. van Dyke," he said, "can you remember your birthday?" Henry responded, and Como wrote it down. "What kind of work did you do?" Henry responded again, softly: "Hardware store."

"Did you feel you had to be more careful than other people in the hardware store? Did you have to check things over and over?" Henry nodded. "More often than other people?" Henry was silent, but eventually shook his head. "Hurts," he whispered, and Kurlan and Como stretched him flat on the bed.

"There's one more thing, Mr. van Dyke," Kurlan said. "We're trying to take small blood samples from all the family members, as I said, to try to find out the cause of this Tourette's syndrome. Wouldja be willing to let us take some blood from ya, Mr. van Dyke? Like as much as they take when ya go to the doctor?"

Henry lay silent for a long moment. The doctors waited.

"You say this is only for research?"

"Just for research," Kurlan replied. "Yes."

Another long silence. Then he nodded. "I guess it would be all right."

Laurie Ernest leaped forward to the rolling table and laid out the supplies: tubes, syringes, swabs. Then she twisted a latex band around Henry's upper arm, and Kurlan lifted a syringe. "Just a prick now," he said quietly, and the first small wave of blood surged into the tube.

The task took several minutes; they needed five tubes of blood, 5 milliliters or roughly a teaspoonful each. Kurlan pulled the syringe away and closed the last of the five tubes with a snap. "Thank you, Mr. van Dyke." He handed it to the nurse. "Oh, look. Your supper's here."

Henry gave the tray a sidelong glance. Kurlan asked him about his children. He offered to take Henry outside, to enjoy what remained of the golden afternoon. Henry shook his head. "Hurts ta sit up." Kurlan made a sympathetic noise in his throat.

"Even in a wheelchair?"

Henry nodded.

"Well, Mr. van Dyke," Kurlan said, "then we don't need to take up any more of your time. We're gonna leave now. I want to thank you very, very much for giving your time, and especially for letting us take some blood. You've been a great help.''

He didn't speak again until they reached the nursing-home lobby. Then he turned to Como. ''Well, we can't say much about that. He's on a drug that suppresses tics.''

Como shook his head and pursed his lips. ''Mellaril and Sinequan,'' he said. ''Terrible drug combination for a person who's sleepy all day. In fact, maybe that's what keeps him sleepy.''

Henry's blood might prove useful later, but only if someone finds a gene responsible for Tourette's. None of information he gave them was new, either. William Holbein well knows that he has TS; in fact, on their way to see his son-in-law's Uncle Henry, he had spoken about the theory that Samuel Johnson, the eighteenth-century English author, also had it.

William's daughter, whom we will call Mildred, also has Tourette's. So does her twelve-year-old son, Russell, who is the *proband* of the study—the first person affected with TS to draw a doctor's attention to this large kindred, and the person from whom relationships are traced.

If Henry had had visible tics, that could have been significant information. He did not, so the visit to the nursing home was a disappointment. But not a grave one. It was not, after all, a momentous event. It was an interlude, a brief respite from the mind-numbing routine necessary to try to put another landmark on the human genetic map.

As soon as the researchers returned to Mildred's house, their center of operations for this field trip, Mildred asked whether Uncle Henry had cooperated. Kurlan was careful to sound jubilant when he replied. He knew she would be very pleased.

Mildred wanted the study to succeed. Just as fervently, she wanted her relatives to accommodate the experts who had come so far for her sake, and who were casting her personal problem in a dignified new light.

Tourette's was relentless, it was embarrassing, it was a monkey on her back. For a long time before she knew what it was, Tourette's made her think she was crazy—crazy as in unsound, irreparably flawed, useless. When her son began to twitch at the age of four, she says, it broke her heart.

Roger Kurlan is a neurologist in his mid-thirties, a soft-spoken and somewhat distant man. He has a beard and mustache, a receding hairline, and the face of a lad. His colleague, neuro-psychologist Peter Como, is handsome and dark-haired. He brings to mind a soap-opera star, not Sigmund Freud. He has a gentle manner and a perpetual look of concern about his eyes; you're inclined to trust him.

Their companion, Laurie Ernest, is bright-eyed, chipper, attractive. She's a nursing supervisor with broad experience in neurology. The visit to Holland was her first field trip.

Routine work or not, they were having as much fun as kids on a bus tour: a different setting, new people, restaurant meals.

Early the previous morning, the three landed at the Grand Rapids airport and collected their luggage, which included an empty Styrofoam container for blood samples and a large cardboard box containing blood-drawing equipment and Tootsie Rolls to reward the children. Another carton was full of printed documents: consent forms, psychomotor test sheets, diagnostic assessment forms, pamphlets about TS. A major purpose of the study was to try out new diagnostic forms for the Tourette's Syndrome Association, to unify the conflicting definitions of TS used in different family studies. On such a dull, procedural matter might hang the discovery of a gene.

Kurlan rented a station wagon, and after checking into a motel drove by memory to the van Dyke house just outside Holland. "Yup, this is the house," he said as they pulled into the driveway. "I remember it. Neat as ever."

He paused before pulling the keys out of the ignition. Everyone in the car was silent, contemplating the front of the house. "But look at all those leaves on the driveway," Peter Como said at last. There was a good two hours' accumulation, considering the canopy of trees and the time of year. "Yes, but they'll be gone by evening," Kurlan said. And they were.

Another visitor might say that Mildred van Dyke is a scrupulous housekeeper, even fussy. Como, Kurlan, and Mildred herself share the opinion that she is pathologically compulsive.

Family studies of Tourette's syndrome show that, while men with Tourette's have tics and other movement disorders, among women it often emerges as a disorder of obsession and compulsion.

In a place like Holland, compulsive housekeeping is an appropriate kind of malady, indeed an almost godly one. To judge from the other houses in her neighborhood, it might be contagious, not hereditary. Certainly no one here would criticize a wife and mother for being too neat.

"Welcome back!" Mildred said at the door. "We might as well go into the family room." Unless you saw her tic, you'd never look twice at her in a shopping mall. She's a casual woman of a generous size—a typical midwestern housewife, you'd think, and a jovial one.

The family room is located off a spotless kitchen, and dominated by a fireplace (two untouched birch logs here, with peeling parchment) and a wall of bookshelves. Mildred van Dyke sees a horizontal wood surface as a mandate for a curio. Besides books and framed photographs, she has displayed here two brass birds, a small doll sitting on an equally small park bench, a plant, a brass jug, four wooden soldiers, two brass loons, a glass plate with a ship etched on it, two pewter goblets, three more wooden soldiers, and a brass basket. There's a great deal to dust. Still, everything shines.

The coffee table was bare this day, except for a recent issue of *Life* magazine. It contained an article about Kurlan's last field trip, to study another kindred with Tourette's in a small town in northern Canada. Kurlan remarked on the magazine.

"Yeah, I keep it to show my relatives and friends," Mildred explained. "I got another one for my mom, to show people. I saw myself on every page of that article." The words tumbled out, as though she had four minutes to talk to him, not four days.

"How'd you feel about that article, as somebody who has the syndrome?"

"Why, it never bothered me. I know some people in my family wouldn't like it at all. Like my uncle Len. He would never admit it's in the family."

"Were you actually in the article?"

"No, no. I just mean I could see the symptoms there. . . .

I remember when I first had Tourette's I was so ashamed about it. Why was I ashamed? I dunno. If I had another disease, like cancer, I wouldn't mind telling anybody about it. But with this— well, for a long time, I just thought I was going crazy. Nobody could tell me what was wrong. I got to the stage where I nearly checked myself in to Pine Ridge. That's over there in Grand Rapids. It's like a mental home.''

All the while, nearly once a sentence, Mildred had been blinking—not an ordinary gentle brush of the eyelashes but a convulsive squint which involved nearly her whole face: nose, cheeks, forehead. Anyone would notice it immediately; Mildred ignores it, as a lame man ignores a limp.

''Doctors would always ask me about the blinking. They didn't know what it was, either,'' she went on. ''I remember this one family doctor said, If you ever want to see anybody about it, I can give you a name. So finally I asked him for the name. I went to this other doctor, and I had been talking to him for about fifteen minutes before I realized what he was—a psychiatrist.'' Mildred stopped seeing him after two visits. She couldn't see how talking about her private life would help her blinking problem.

''What about your son?'' Roger asks. ''Do the other kids sometimes ask him what's wrong?''

''Yeah, sometimes the other kids ask him why he blinks so much. He tells them he has Tourette's syndrome. I gave him some real long words to use in the answer, and that generally shuts them up. I also gave each of his new teachers a pamphlet, and that has made a big difference. They're all really understanding, once they read the pamphlet.

''In fact, if my son had to have Tourette's I'm real glad I have it too. At least I understand. If I didn't have it, I'd be like my husband. I'd say, 'Can't you stop that? You're driving me crazy.' I tell him—here she whispers—'Don't say anything. He can't.' ''

Even as easygoing and diplomatic a man as Mr. van Dyke, evidently, has trouble ignoring his son's mannerisms. Tourette's is not a fatal condition, merely a relentless one.

''How's your mother?'' Roger asked, and pulled out his photocopy of the pedigree to begin recording new deaths and births. ''So who do you think we'll see this evening?''

This was a matter of great concern to Mildred. She was nervous that none of the distant relatives she had called in the past few weeks would understand or care enough to come. They might stay away out of fear of the blood-drawing, or out of fear of being associated with her problem, or, worse yet, of being told that they or their children had it.

Kurlan asked about one of her brothers.

"It's always so strange," Mildred mused. "Why it should pass down to everyone in my family—I mean my father and me, then Russell, but not in his?"

"Well, to some degree it's just chance," he replied. "Fifty-fifty, like a flip of the coin."

"You're the doctor. You'd know," she granted. "But couldn't it be that it's because I have it on both sides of my family?"

"Could be," Kurlan agreed, distractedly. He was attending to her on two levels, listening to her words but also watching the way she touched her face, again and again, between the mouth and nose, probed the bridge of her nose, fingered her chin, plucked at the front of her sweater. Touretters seem to revel in sensation, the sound of sounds, the smoothness or roughness or softness of things, the rhythm of repetitions. As she searched for a magazine in the rack, Mildred touched each one in the upper left corner before flipping past it.

Later, when the researchers asked to move their supplies to the basement rooms where they would work, she apologized. It's probably a disaster down there, she warned: dust, cobwebs, the mess from her son's Legos. God knows. The Lego "mess" consisted of three figures, displayed as though the billiard table were a gallery floor. As for cobwebs, a spider wouldn't have time to finish the job before one was swept away.

Mildred used to spend nearly half her time cleaning house. Each day, she would remake the children's beds whether or not they needed it, and make sure that Stu's shirts were hanging assorted by color and the socks sorted and lined up in the drawers. Each week she would dust the top of the doorframes. One day, she looked at the dustcloth in her hand. It was clean.

"This is crazy!" she said to herself. "Why am I doing this?" By now she has the housecleaning down to two days a week.

■   ■   ■   ■

Near the billiard table, Laurie Ernest set out the bloodletting supplies on a card table and on the bar. Scientists in this field sometimes call these "Dracula parties." To the subjects, they're family reunions.

In a few hours, distant relatives would begin to trickle in. For the next three days, there would always be a small crowd in the family room—always a different crowd, except for Mildred van Dyke in her beige recliner, always the smell of coffee, the flicker of the World Series running silently on the TV, and the warm sound of conversation.

" 'Ja sell all that honey yet?"

"All gone."

"All at the farmers market?"

"All six hundred pounds. All gone. The little market in town wanted us to supply their store, but it was all gone."

"How'd ya separate it from the comb?"

"By centrifugal force. Then we screen it."

"Are you doing your homework?" Mildred called to her son.

"I'm in the process of thinking seriously!" he called back, from the kitchen table. A handsome blond boy of twelve, he leaned over his notebook, frowning. Every few moments, in the process of thinking seriously, his head bobbed up and down or lurched to the left, toward his shoulder. His gaze returned instantly to the page. Just as often he gave off what sounded like a belch.

Mildred's sister-in-law was reading the article in *Life* magazine. "It says here the disease is rare! I don't think it's rare, not considering all the people I know who have it. And not just family! The man who fixes our refrigerator. He's always making noises. *Pfffffft. Pfffffft.*"

One of the most serious and bothersome symptoms of Tourette's syndrome is known as *coprolalia*, the involuntary and repeated utterance of obscene or objectionable words. Often, according to Roger Kurlan, someone with coprolalia may not say whole words but just the initial sound: "f——, f——, f——." There's also a mental form, in which a person repeatedly thinks obscene words but does not speak them aloud.

"Other conditions can have tics," Kurlan adds. "People

with Huntington's disease can have tics. But virtually no other conditions show coprolalia.''

The textbooks say that 30 percent of people with Tourette's suffer coprolalia at one time or another, although the figure may be much lower. The textbook definition of Tourette's has been under considerable revision. Today simple vocal tics, inarticulate sounds such as whoops or grunts or belches, are also included as symptoms of Tourette's syndrome. Conservative estimates are that some 100,000 Americans are affected by Tourette's syndrome today, although the real incidence is hard to assess without a genetic test.

In 1979, Dr. Mary Anne Guggenheim of the University of Colorado Medical Center reported a five-generation kindred in which 17 of 43 members appeared to have Tourette's syndrome. Throughout the early 1980s two teams of researchers, from Yale University and from the City of Hope Medical Center in California, separately began to publish family studies of the syndrome. Computer analysis of the pedigrees fit with the theory that Tourette's syndrome is a dominant, single-gene disorder involving coprolalia, tics, obsessive-compulsive symptoms, and—depending on whom you believe—perhaps much, much more.

Until five years before his second trip to Holland, Michigan, Roger Kurlan had no particular interest in Tourette's. He saw it as a rare syndrome he would be unlikely to encounter often in his practice. Then, one afternoon in November 1983, a lanky, balding man appeared in the neurology department at Strong Memorial Hospital in Rochester, New York, without any medical insurance and without an appointment. He asked to see Kurlan's boss, Dr. Ira Shoulson, an internationally renowned expert on movement disorders.

The secretary asked him to identify himself. He gazed back at her good-naturedly, blue-gray eyes set within a dozen wrinkles, and identified himself as David Janzen, thirty-eight, from Alberta, Canada. She took no special note of this, apparently having no concept of how far he had traveled at his own expense to reach her desk. In a neurology department, she would not even have been startled if Janzen had suddenly had one of his tics, shrugging and twisting his arm upward as if to protect himself.

As Mildred van Dyke once had, David Janzen thought his convulsive motions and whoops meant he was going crazy. So did many other people, unless they sized him up instead as a drug addict. Often, during his extensive travels in search of a reason for his condition, strangers would ask him for drugs because they thought he was addicted.

In fact, it's hard even to get a drink in Janzen's remote hometown. La Crete, Alberta, is the center of a farming community dominated by a group of Mennonites, members of a conservative Protestant sect.

Janzen, whose tics were severe enough to be disabling, had spent nine years searching for an explanation. He had come all the way to Rochester to verify two diagnoses he had been given nearer home. One was unsatisfying and the other terrifying. Doctors either told him he had myoclonus, which is a descriptive term for paroxysmal movements, or that he had Huntington's chorea, a degenerative disease of the nervous system that leads to dementia and early death. Dr. Shoulson, he had heard, was the man to see about Huntington's disease.

At the time, Kurlan was the research fellow in Shoulson's department, the person assigned to see all new patients and report about them. As such, he was the first to identify David Janzen's problem. "You don't have Huntington's disease," he told him immediately. "You have Tourette's syndrome."

Kurlan asked if anyone else in Janzen's family had the same symptoms. Sure, he replied, and began to list them. Kurlan hardly expected a list, but Janzen named about a dozen family members who had tics and some, including David himself, who had also had coprolalia.

Kurlan decided to admit Janzen to the hospital for observation and treatment. He had a colleague draw up a preliminary pedigree, and he asked Janzen whether he could return to Rochester with other members of his family. A few days later, at Janzen's bedside, Kurlan took back the request. "I don't think you'll need to come back to us," he said. "We'll come to you."

That very month, as it happened, Shoulson and thirteen coauthors from various institutions had reported another triumph of reverse genetics: they had located the gene for Huntington's disease on chromosome 4, near an "anonymous" DNA fragment, a segment that had been generated using restriction enzymes but was not a gene and had no function as far as anyone knew. The reasons for Huntington's chorea were a total mystery; the location of the presumed gene near that anonymous fragment was the first clue to its solution—or so everyone took for granted.

"It is likely that Huntington's disease is only the first of many hereditary . . . diseases for which a DNA marker will provide the initial indication of chromosomal location of the gene defect," the research team wrote in the journal *Nature*. They were right about that.

The kindred involved in that landmark study was an interrelated community of Indians living along the shores of Lake Maracaibo in Venezuela. It had been a tour de force of medical fieldwork quite different from Kurlan's later visits to Holland, Michigan. The Huntington's team had spent a month in Venezuela each year for three years, living in a house on stilts in Lake Maracaibo, running open-air clinics, and paddling blood samples downstream before flying them to the United States for analysis.

David Janzen materialized in Rochester only weeks after Shoulson had returned from his latest trip to Venezuela. "Everyone was real excited about the linkage studies," Kurlan recalls. "I guess literally the second thing in my mind, after I'd diagnosed [Janzen's] condition was, Boy, this would be a heck of a family for a linkage study. If he's right."

"I was a little incredulous that so many people were affected," he admits. Given the diagnostic blunders that had sent David Janzen to Rochester, Kurlan doubted he could verify Janzen's story about his relatives without seeing them in person. "But basically, we just said, Hey, if they can do it in Venezuela with Huntington's disease, we can certainly do it in Canada with TS."

Kurlan contacted the coordinator of the Huntington's disease study, psychologist Nancy Wexler, and asked if she knew any molecular geneticists working on TS. She mentioned the name of Kenneth Kidd at Yale University, who was interested in the genetics of psychiatric diseases. Within days of meeting

Janzen, Kurlan called Kidd, told him about the kindred, and asked if he was interested. The response was immediate: Kidd had been heavily involved in family studies of Tourette's at Yale, and was looking for families of about fifteen living members with two individuals affected with TS. If they were real, the Janzens would be far better than that.

Next Kurlan called the Tourette's Syndrome Association in New York City. "Look, I've got a really hot thing here," he told a TSA official over the phone. "I think we have found the largest family in the world with Tourette's syndrome. I've made contact with a molecular geneticist at Yale, and he's interested. But we really need to go up to Canada and study this family."

The deadline for grant applications had passed the week before, he was told. But get your application in right away, the TSA official responded. Take as long as it takes, but do it. We'll wait.

The following summer, the association gambled $14,000 on David Janzen's reliability. Kurlan and two co-workers traveled to northern Alberta to meet David Janzen's relatives and try to obtain some of their blood. At the moment of his arrival in La Crete, Kurlan began to trust his patient's story. Almost the first person he saw in the small town had noticeable body tics.

The trip was a huge success. Dr. Kurlan met sixty-nine people, took forty-four blood samples, and confirmed that about a dozen of them seemed to have Tourette's syndrome. In subsequent summers Kurlan and Como visited two other Canadian communities where Janzen family members lived, and in 1987 they returned to La Crete. That follow-up visit, and David Janzen's story, were the subject of the article in the *Life* magazine on Mildred van Dyke's coffee table.

By then the study had expanded to include five large families—kindreds in Oregon, Boston, and Kansas as well as the Janzens and the extended family in Holland, Michigan. It is possible to predict with computer models how many people you need to study to determine a genetic linkage, if you assume a certain inheritance pattern. If, as it seemed likely, the gene would prove to be acting alone and in a dominant fashion, these five families should be enough to pin down its location. The second trip to Holland might clinch it.

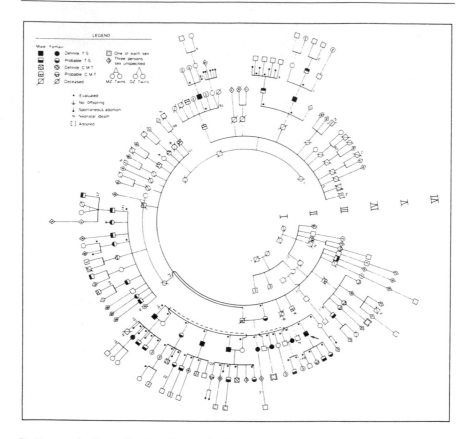

Pedigree of a large family affected by Tourette's syndrome. (From Roger Kurlan, Jill Behr et al., ''Familial Tourette's Syndrome: Report of a Large Pedigree and Potential for Linkage Analysis,'' *Neurology* 36, June 1986, p. 773; with permission)

■   ■   ■   ■

Kurlan and Kidd are on dangerous ground, and everyone knows it. There are perils as old as human history in this practice of classifying peculiarities among various people. That's exactly the business that human genetics is about. At times, it has been a bad business—or, more precisely, a bad science.

For most of this century, human genetics has lived under the shadow of its early excesses in the name of eugenics. Eugenics was not just Hitler's foible. It began earlier, with a British scientist, and spread quickly to America. We still have its legacy in some of our immigration laws.

The movement began with Sir Francis Galton, a cousin of Charles Darwin, who produced statistical studies on the inheritance of all sorts of ill-defined human qualities. Based on his observations about the aristocracy, for instance, he proposed a genetic trait for "eminence." Galton went on to campaign aggressively for active efforts to breed the best into humanity and restrict the breeding of the worst.

In general, the eugenicists who followed Galton tended to neglect the importance of cultural kinds of inheritance, and to be blind to their own value judgments and class prejudice. As one geneticist has pointed out, none of them ever proposed sterilizing European royalty because hemophilia ran in the family.

In the United States, Galton's disciple Charles Davenport, a zoologist, established America's first genetics research center, which was called a "Station for Experimental Evolution," at Cold Spring Harbor on Long Island in 1905. There he set up a Eugenics Record Office to keep statistics on Americans and to advise individuals and society how best to breed.

The critical problem, as University of Michigan historian John Higham wrote, was that the eugenicists "made virtually every symptom of social disorganization look like an inherited trait." Consider the advice from a committee of eminent geneticists and lawyers to the Eugenics Committee of the United States of America, written in 1921.

They recommended "selective immigration by refusing immigrants below the determined standard of excellence" so that "defective heredity and disease can be kept out" and also "elimination of defectives . . . by segregation and, in extreme cases, sterilization. In this way the number of insane, feeble-minded, epileptic, criminalistic, and paupers can be greatly reduced in a generation or two.'' Part of the tremendous fear and indignity inside Ellis Island was due to eugenics.

Whatever the motives behind it, the eugenics of the early twentieth century was probably doomed in a place such as the United States. As Harvard biologist Ernst Mayr points out, ''there is a complete ideological clash between the concepts of egalitarianism and eugenics.''

The effort eventually foundered on two problems: a lack of good tools to study inheritance, and the definitions of ''best'' and ''worst.'' But the science of human genetics, which gave rise to eugenics, has hardly withered since the demise of eugenics. We have the tools now. We're left to grapple with those definitions.

Perhaps the saving grace of the present is that we know enough to know how little we really understand about human inheritance. Perhaps not.

What does this have to do with Tourette's syndrome? In barely more than a decade, studies have transformed its image. Once a rare behavioral disturbance of ''street crazies'' who shout obscenities, today it is a more common neurological disorder of twitches and compulsions. Now there are suggestions in high places that the syndrome may be very common, and may in fact encompass many ''symptoms of social disorganization.'' Now people know how to look for the hypothetical gene, and are doing so.

A week before Kurlan and his two colleagues returned to Michigan, Tourette's took center stage at the thirty-ninth annual meeting of the American Society of Human Genetics. For the most part, such meetings are dispassionate and often boring recitals of scientific esoterica. But they are also the major place where the battles and rivalries inside the cloisters of research can burst into the open.

On October 13, 1988, Dr. David Comings, feeling some apprehension, strode toward the podium in a ballroom at the Hyatt Regency Hotel in New Orleans, to deliver his presidential address. Comings took the opportunity to engage his listeners' sympathy.

''This morning I want to discuss the subject of the genetics of human behavior and its implications . . . ,'' he began. ''Initially, the thought of presenting this subject gave me a few twinges of trepidation, given the nightmare that befell one of our members when he entered into this territory over ten years ago.''

He did not elaborate. For the well informed, he was invoking the memory of geneticist Park Gerald, codirector of the Harvard Newborn Study. That study was disbanded due to unceasing harassment, including threats to the safety of the researchers. What generated such an extreme protest? An attempt to analyze chromosomes in the blood cells of all the babies born at the Boston Hospital for Women.

The Harvard researchers had been trying to test the assertion that boys who possessed the XYY karyotype (an additional Y chromosome beyond the usual male set of X and Y) tended to become deviant, delinquent, even criminal. That correlation ''has received more unfortunate publicity, more speculation, and has led to more controversy than any other single discovery in human genetics,'' declared the man who was president of the genetics society at the time. Because of it, the Y chromosome had been named the ''criminal chromosome,'' and the XYY genetic type had even been used (unsuccessfully) as a defense in several murder trials.

According to Gerald's co-worker, psychiatrist Stanley Walzer, XYY boys have normal and even high IQs, but are likely to suffer speech and reading problems. The study was designed to determine the full effects of the extra Y chromosome, and to resolve the nature-versus-nurture aspects of the issue, by catching such boys at birth.

Critics of the study charged that it was not only worthless but might be positively harmful to those babies branded at birth as deviant. XYY might be nothing more than a marker of ''adverse social factors in the parents' background'' that led to an increased risk of chromosome abnormalities, they said, adding their opinion that the studies were ''ideologically influenced.''

Although the studies had been approved by a review board at the National Institutes of Health and by a special vote of the Harvard faculty, Gerald and Walzer suspended them in 1975. At the time, Walzer told a journalist that he felt ''persecuted and drained.''

By recalling the story, Comings signaled to the audience that he too was feeling embattled. No one had suggested that he suspend his research, but the appointed chief of America's human geneticists certainly had his critics. As many of his listeners knew, one of them was scheduled to challenge Comings that very afternoon.

Continuing his address, Comings began to reminisce about his own research, into Tourette's syndrome, at the City of Hope Medical Center. "When Brenda [his wife, a psychologist] and I first became involved in studying this disorder we thought, as did everyone else, it was an extraordinarily rare disease of the type that is of interest only to medical geneticists." Then, at their bidding, Tourette's patients began to come in to the clinic, he said, one or two a month.

The Comingses had been featured on one segment of a local television talk show aired only in southern California, during which they spoke about Tourette's and gave out their phone number. Afterward they went out to dinner. When they returned home, more than sixty telephone messages were waiting for them.

They launched into a busy schedule of speaking about Tourette's at local schools. Every visit seemed to call forth one or two new cases. Eventually the Comingses began to collaborate with school psychologists, to seek out children with Tourette's. By the time of the address, he said he had located 1,200 children, plus 500 with "related disorders."

The studies that followed, Comings told his audience, involved "the kind of compulsive pedigree taking that only a dyed-in-the-wool geneticist can do and the kind of emotional probing that a clinical psychologist can add." (He did not remind them that he is also a physician.)

First the Comingses focused only on the tics and vocalizations they noticed. Then, as their experience grew, they began to watch for obsessive-compulsive disorder and later for attention-deficit disorder, an inability to concentrate that has been linked to hyperactivity. More recently, they have expanded their observations about Tourette's patients to include conduct disorders (such as vandalism, fighting, hurting animals, and the like), learning problems, panic attacks, depression, and mood swings. In the last two years they have added alcoholism, drug abuse, and obesity to the long list of behaviors potentially associated with Tourette's.

"As we continued to take pedigree after pedigree it began to dawn on us that they all made more sense if most moderate-to-severe cases were homozygous for the TS gene," Dr. Comings told his fellow geneticists. "But other members of the family who were carrying a single gene could also express it as one or more of the aspects of the spectrum of behaviors."

The studies suggest "that the frequency of TS in boys is approximately one in 100 and at least half are homozygotes," he continued. "Our experience suggests that about half of the carriers have some symptoms of the TS spectrum, or about 6 percent of the population." He meant 6 percent of the *entire* male population, not merely of the study's subjects.

Comings flashed onto the projection screen the pedigrees of several families he had observed. The first began with a sixteen-year-old alcoholic and drug abuser who had been in a "residential treatment facility" for over a year. His alcoholic father had chronic motor tics; his aunt was alcoholic.

In the second family Comings showed, a son with Tourette's had a father with tics; an alcoholic, drug-addicted aunt with attention-deficit disorder and conduct disorder; an emotionally disturbed cousin who was a drug addict; an alcoholic uncle and another uncle with "full-blown" Tourette's syndrome, attention disorder, drug addiction, and alcoholism; another alcoholic aunt and a third aunt who had problems with drug addiction and weighed more than three hundred pounds.

The litany trailed on and on, pedigree after pedigree, resonating with images in the mind: rusting cars in front yards, loud radios, unkempt hair, ill-mannered scenes in public places. Of course, Comings did not even hint at such images—or, certainly, at the image of impeccable suburban houses where blond boys quietly do their homework. A scientist does not deal in images, after all, but in findings.

"To summarize these ideas," Comings concluded, "we would suggest that a certain proportion of people in the population carry this gene, perhaps around 13 percent, and that it may be expressed as . . . a wide range of addictive, compulsive, mood, and learning disabilities." Comings ended with a call for more research alliances like his own, genetics with behavior. He was greeted with loud applause.

None of what he had said was surprising to anyone who read the monthly journal of the society sponsoring the meeting. About a year earlier, the Comingses had revealed their findings in a remarkable scientific report: six articles, back to back, in the *American Journal of Human Genetics*. Their articles dominated the entire issue. Comings had just stepped down as editor, and was to become president of the society.

The concluding paper in the series was a lengthy neurological analysis, in which Comings hinted at a possible brain defect common to the many phenomena he and his wife had seen in their study subjects. Among other things, he extensively reviewed prior animal studies in which induced brain damage caused behaviors such as sniffing, licking, and biting, repetitive head and limb movements, and abnormal sexual activity.

He also published a table linking such things as sleep disturbance, poor response to stress, inability to take responsibility for one's own actions, paranoia, and inappropriate sexual behavior, all listed under the column "TS," with similar behaviors found in studies of brain-damaged animals. To Comings, it all sounded like a specific disorder of the limbic system, which is an ancient part of the human brain, the seat of emotions, sensations, and appetites.

"The limbic system has been characterized as controlling the four Fs," Comings concluded, "fight, flight, feeding, and sexual activity [sic]. It has not escaped my attention that the reason many of the disorders described in the present series of papers are so common is that they are (1) genetic, (2) dominant, and (3) result in disinhibition, especially of sexual activity."

What more perfect way for a gene to survive? It could induce its own reproduction by promoting promiscuity.

That issue of the *American Journal of Human Genetics* landed like a bomb at the Yale Child Study Center, where geneticist David Pauls was just completing a seven-year family study of Tourette's syndrome in collaboration with Kenneth Kidd. Mothers in the study began to call as the Comings reports gained publicity, to ask if their children were doomed to suffer schizophrenia or to sexual deviance, as well as enduring tics.

A bizarre problem that Pauls and his team had been describing as comfortingly neurological threatened to slip back into the murky world of psychosis. Pauls and Kidd began to draft a rebuttal to the Comings' articles. It appeared several months later in the same journal, as a letter to the editor.

The lengthy letter included five specific criticisms of the Comings' methods: They had not used a suitable control group, Pauls and Kidd charged. They had relied on questionnaires sent through the mail, seldom on personal interviews. They had failed to take account of such matters as social class, and had tipped the scales in favor of severely disturbed families by the way they sought out their subjects. Pauls and Kidd also took the Comingses to task for speaking concretely and in detail about the inheritance of a gene whose existence had yet to be proved.

Kidd and Pauls saved their most forceful language for the Comings' conclusions about the "four Fs." They wrote that "the authors present no data to demonstrate that individuals with TS are sexually disinhibited. . . . To attach such a label to individuals who have already suffered tremendously because of their illness is at best insensitive; to do so without having any data to substantiate the claim is inexcusable."

In New Orleans, David Pauls was scheduled to present some data at a concurrent slide session, a far less formal event than a presidential address.

By the time Pauls began to speak, the room was full. The title of his remarks was "Family Study of Tourette Syndrome: Evidence Against the Hypothesis of Association with a Wide Range of Psychiatric Phenotypes." Hardly a banner headline, but on the page of the 1988 meeting schedule it stood out as eloquently as a thumb to the nose.

David Pauls is slender, bearded, professorial. He is a mathematician with a Ph.D. in genetics, a methodical person whose own research habits are, he admits, somewhat compulsive. His ancestry, as it happens, is Mennonite. Pauls began dispassionately; he might have been talking about turbulence patterns in teakettle steam.

"However one sees a TS patient, they are struck by the wide range of behavioral and psychiatric and other manifestations that appear to be there," he said. Pauls went on to describe the textbook definition of Tourette's, and to say why his own studies support obsessive-compulsive disorder as part of the syndrome.

"Look at the definition for obsession," he said. "It's recurrent, persistent ideas, thoughts, images, or impulses that are not experienced as voluntarily produced. Sounds very much like mental tics. Compulsions are repetitive and seemingly purposeful behaviors that are done according to a certain rule or in a stereotyped fashion. It becomes very difficult at times to distinguish between complex motor tics and compulsions."

In his study of Tourette's families, Pauls assessed people using detailed interviews that lasted nearly two hours for a family member who seemed normal, and from three to six hours for a person who appeared to have the syndrome. "We try to document very carefully the presence or absence of any behaviors," Pauls told the audience.

With such methods, he said, the Yale team saw an "incredible increase" in the rate of obsessive-compulsive disorder in Tourette's families, as compared to families without Tourette's syndrome or families of children with Tourette's who had been adopted. This was true whether they traced the Tourette's families starting from people who had tics only, or from others who had obsessions and compulsions as well as tics.

"As we heard this morning," he went on, "it's been suggested, primarily by Drs. David and Brenda Comings . . . that there were a number of other illnesses that may in fact be associated." He went on to list the proposed "variant" expressions of the hypothetical TS gene: attention-deficit disorder, conduct disorder, learning disability, language disorders, phobic disorders, panic disorders, schizoid behaviors, major depressive disorder, manic depression, as well as sleep disturbances, alcoholism, drug abuse, and obesity.

As he had done for obsessive-compulsive disorder, Pauls went on, he reanalyzed his data focusing on two of these other traits: attention-deficit disorder and manic depression. At first glance they did appear to be "marginally" more common among Tourette's families than among other families, he said. But if he looked to see which diagnosis came first, these problems were more common only when they came after the diagnosis of tics or other classic TS symptoms. The implication was clear: attention problems or depression arose as a result of having to cope with TS, not because they were intrinsic to the syndrome.

By the time Pauls finished speaking, David Comings was at a microphone in the middle of the aisle, ready to step into the closest equivalent the science community has to a wrestling match: the question-and-answer period. Many people in the audience turned to look at him. A large man with graying hair rather longer than average, Comings gazed levelly at his adversary. He spoke in a jocular tone, like a talk-show host.

"Given the letter that you wrote to the journal, . . . your statement here was in fact rather remarkable. You said, 'The frequency of the other neuropsychiatric disorders in our proband sample were comparable to those reported by Comings and Comings.' I assume from this that you're saying that, when you looked at the Tourette's syndrome patients, many of these other disorders were in fact present."

"The people who come into our clinic do in fact have other behavioral and psychiatric diagnoses," Pauls responded. "They do in fact have learning disabilities, they do—"

Comings broke in. "Because just three months ago you were claiming that that absolutely, unequivocally was not the case."

Pauls replied: "No, that's not what we said in the letter. We said that our data do not support the hypothesis that there is an increase in the relatives. We did not say that these patients that are seen in our clinics don't have these other diagnoses." People with more than one problem might be more likely to come to clinics, he argued. If behavior problems such as conduct disorders were really intrinsic to Tourette's, he added, they should also be increased in these families, of themselves. "Our data simply don't show that," he concluded.

"Our data do show that," Comings replied. "Did you look at alcoholism?"

"We have looked at alcoholism and it's not significantly increased, above population prevalences."

"Were these people from Canada as opposed to the United States?"

"These were all people from the Northeast," Pauls replied. Comings muttered something under his breath, and there was laughter from the center of the room.

As Comings turned away, his wife, the psychologist, advanced to the microphone. "Um, Dr. Pauls," she began, "do you see patients?" In a room full of geneticists, who lean to denim and sneakers, her appearance was as defiant as her words: designer suit and bright turtleneck, tight chignon, long red nails.

"Yes," Pauls replied. "I'm not a clinician, but I have seen a number of individuals in these families."

"It's rather interesting to me that in the context of your speech you made a statement that if there was just chronic motor and vocal tics, it would take maybe one or two hours to have an interview, and it would take three to six hours if there's a pathology involved. Well, if there's no pathology involved, why does it take so long?"

"Because we ask an enormous amount of questions," Pauls replied, and there the chairman of the session closed the discussion.

But it continued just outside the door, in the broad corridor behind the ballroom. The Comingses had been cornered by Arno Motulsky, from the University of Washington in Seattle, one of the pioneers of human genetic studies. A small knot of medical reporters listened.

"So what do you think?" Comings the geneticist asked Motulsky, with a smile.

Looking at a reporter, Motulsky responded: "You want me to say so in front of him?"

"Yeah, go ahead."

"I think you're making claims that hearken back to the early days of genetics. I disagree with you completely. You make these grandiose schemes without evidence. You revolutionize genetics if it's true, but I think it's very unlikely. I think you're doing genetics a disservice." Motulsky told me later that plenty of people are saying that behavioral genetics is racist, sexist, classist, any other "ist" you can think of, and that he felt the Comingses were adding to the problem by defining their studies too loosely.

"But you can make anything a single gene," Motulsky was insisting, "if you define it as dominant and then trace it back like this." Brenda rolled her eyes heavenward and smiled indulgently at a reporter. "Interview over," she said.

Across the broad corridor, leaning against a pillar, David Pauls was watching. "It's . . . it's difficult," he mused. "It's like when he's talking, I miss something. It's like we're not speaking the same language. Like when I'm speaking, he doesn't understand what I'm saying. Or doesn't hear me.''

They are not, in fact, speaking the same language. The debate between the two Davids is just another skirmish in an ancient war between clinicians and academic scientists, a battle between massive clinical experience and observation, on one side, and meticulous study design and high-powered statistics on the other.

The Comingses continue to find an association based on controversial methods. Pauls fails to confirm it—a negative result, a mere vacuum of evidence, which, in the arena of scientific conquest, is rather like firing with blanks. Awaiting a genetic test to settle the matter, the adversaries will continue to raise dust.

Gradually, as Mildred van Dyke sat in the beige recliner welcoming relatives into the family room, morsels of gossip dropped here and there. A cousin who was always depressed has committed suicide. A distant nephew may be illegitimate. So-and-so drinks too much.

Sometimes a member of the research team was present to hear. He might listen gravely or even raise an eyebrow, but made no effort to record the information. Except for illegitimacy, none of this is relevant to the study at hand. Kurlan and Como side with their research collaborator, David Pauls. They feel that studies of TS families who never seek help in a clinic will lead most reliably to the true definition of the disease.

Over the course of four days, the team got to know thirty-three more of the van Dykes and Holbeins. To an untrained eye, the family would appear to be a pleasant, quite normal group of people. To a psychologist and neurologist attuned to the subtle signs of Tourette's, the family seemed to be deeply penetrated by the syndrome.

Almost none of the relatives had Mildred's convulsive blinks, but the number who had subtle evidence of tics or compulsions began to be disturbing. What was needed was a clear pattern for David Pauls to compare with the DNA polymorphisms in the blood cells. Too *few* normal family members is as much of a problem as too many.

At the outset, they had delighted in catching clues of TS. "Wait till you see him," Como had said after assessing the first visitor of the week. "Classic, classic compulsions, and every sort of tics."

"Every one of those sibs is affected," Kurlan replied. "You say he has tics?"

"Yeah, he had the most wonderful truncal tics, and the classic rituals. Wait till he tells you about his nose."

Como had been observing as a huge man in overalls completed a few simple but mind-engaging tasks, such as crossing out every number 7 that appears in line after line of random digits. Think of him as Luke.

Luke scowled and breathed heavily as he worked in silence, behind a closed door in a basement room. His task now was called trailmaking; Como had asked him to trace an ordered path among the letters and numbers littered across the page, from A to 1 and B to 2 and C to 3 and so on. He clutched a pencil in his mittlike hand and guided it across the paper.

While Luke was busy and looking down, Como's eyes were dancing, roving from Luke's face to his hands and his back and chest. Body tics may quiver across the trunk, a symptom no novice to Tourette's syndrome would ever see. After a minute, Como leaned back quietly, to look under the table for movements in Luke's feet or legs.

If the pencil-and-paper tests are a little irritating for the subjects, observing them is child's play for a trained neuropsychologist. Como has done this scores of times, hundreds, in the last few years. On the usual day in Rochester, he would be evaluating head-trauma victims and people with Alzheimer's disease or other forms of dementia. Giving a test to giggling teenagers and strapping laborers who have taken a day off is hardly a challenge. An intern could do that part; even a child could. But few people have as much training as Como has in recognizing the subtle signs of Tourette's. So he sat at the card table, and watched, for four days.

Luke had owned up to a history of Tourette's syndrome. "Mildred's father has it too," he told Como. "He and I both had it when we were young. I must have had more willpower. But he's always pullin' the neck like this. To cover it up, he puts his hand on his chest."

Luke also admitted to certain harmless compulsions that showed up in his work as a carpenter. "When I hammer a nail," he said. "Da da da . . . Da da da . . . You continue like that, till you're done, always the same number."

"Do you still do any of these . . . these rituals?" Peter inquired.

"I thought of that this morning. I wouldn't say so outside here [he looks at the door], but I still get my neck—my neck. . . . When I was still workin', you know, I was always rubbin' my nose. People used to tease me, say I was always the one with the black under my nose. Because I would rub my nose like this [back, forth] and my hands was dirty. . . ."

The large man's breath whistled as he moved on to a second test, translating letters of the alphabet into symbols using a key at the top of the page. Como continued to watch, glancing now and again at a stopwatch in his hand. People with tics or obsessions may take longer, because their symptoms may distract them from memorizing the code. After a few minutes, Como clicked the button on the watch and told Luke to stop working.

"Thanks very much. We really appreciate your coming down, 'cause it looks like it's on both sides of the family. So Mildred has a double whammy—as we say in the business."

Upstairs, Kurlan had shut himself into a small spare bedroom with one of Mildred's brothers, Peter. "So you still think you're affected?" he asked, rhetorically. As soon as the man had sat down he blinked violently, compressing the tops and bottoms of his eyes and his nose.

"Think so," he said. "Yeah."

"I can see the blinks. What about noises?"

"I got throat clearing, yeah. I had the cough, and I got rid of the cough, and now I have the throat clearing. I've always had habits. Feels like there's always something in my lungs." And he cleared his throat.

"About the movements: Anything other than your eyes?"

"I have a tendency to pull my right eye toward my right ear. And I tend to pull my nose. Always to the right side. I don't know why." He proceeded to do so, demonstrating: a Rumpelstiltskin sort of movement, like the grimace of an elf.

It was a brief evaluation—only a reassessment, and the diagnosis was obvious. The next person Kurlan saw was Mildred's nephew, a boy we will call Ricky. As soon as Ricky sat down, Kurlan reminded him of the tics that were so obvious in Russell, Mildred's son. Did he ever notice any tics or twitches like that, about himself?

Ricky twisted his mouth slightly, as if he found the question distasteful. He responded quickly: No. No tics, no grunts. And that was so, evidently. Kurlan reached behind himself and picked up a magazine from the daybed. He flipped through it.

"How about doing a little reading for me, Rick? Just start here, please." It was not a children's magazine, and some of the words were long and unfamiliar. Ricky could master it, but only with concentration. In about the third sentence, the upper half of his face contracted in a great blink. He read another sentence, then stopped and looked up.

"Keep going!" said Kurlan, teasingly. "You didn't think you'd get off that easy, did you?"

Ricky labored on. After a few more sentences his head gave an almost imperceptible jolt, the way anyone would follow a fly passing close in his peripheral vision. But there was no fly.

Kurlan let Ricky read to the end of the paragraph, then stopped him and thanked him for coming.

"That's all?" said Ricky.

"That's all," Kurlan said, and Ricky left the room, smiling smugly.

Over the next four days, more than thirty times, the neurologist and the psychologist repeated the same questions and the same tests, always vigilant for any blinks or twitches, or any evidence at all of strange compulsions. Back in Rochester, they would slowly compare notes later, adding information from long questionnaires the subjects would fill out and mail back, and from the videotapes Laurie Ernest was making of everyone. That was easy for everyone; she'd ask people to read briefly from a magazine, and then just to talk into the camera. Eventually, the diagnoses would be coded and matched with the corresponding blood samples.

Downstairs, a woman who had come during the last field trip asked Como whether she had been assessed as having Tourette's. "Not that it matters. I don't really care. I'm just curious, I guess."

"Well, let's see what we wrote last time," said Como, referring to her file from the previous visit. "Says, 'None.'"

"Oh, good, then I'll stop worrying." She asked how close they are to finding the gene.

"I think we're very close—between your family and the family in Canada, we're very close to establishing a link. We just need to send some more blood samples to Yale—but it should really be just a matter of months."

In the family room, no one seemed to talk about the interviews or the funny pencil tests; the chief topic of conversation, when the subject at hand came up, was the bloodletting: Who was afraid to come for that reason. What hefty grown man declined to drive over to Mildred's for fear of turning pale at the sight of a needle?

Laurie Ernest, the nurse, who was the least senior member of the team, in a sense had the most important job: To calm people and draw their blood with the least possible fuss. "It's Dr. Kurlan's policy not to torture the subjects," she joked with one aunt, as she probed her inner arm for a vein.

A surgical nurse by training, not a phlebologist, Ernest was probably more accustomed to putting something into the bloodstream than taking something out. Beside the card table, she would scrutinize the pale side of arm after arm, looking for the right spot to stab. She knew well what could go wrong. There is no road map of a particular arm.

Above all, she knew, she must not upset anyone, must not make the experience so awful that someone would refuse to cooperate with the team later or, even worse, would alienate the rest of the family. One woman had fainted during the last visit, and almost everyone joked about it this time. The last thing she wanted to do was to stab anyone five times.

So on the first day of the trip, she arose at 4:30 A.M. to go along on first rounds at the hospital in Rochester, so that she could practice using a device that allows you to change tubes without removing the needle from the vein. Then she caught the plane to Grand Rapids. By the time she faced her first nervous subject in the van Dykes' basement that evening, Ernest was running on adrenaline.

"Hi!" she greeted him. "I'm Laurie! Are you ready to have

your blood drawn?'' The lanky young man had a guarded look to his eyes. The needle slipped easily into his arm, and as he looked away blood began to gush into the first tube. When she eased it away to replace it with another tube, crimson blood followed, spurting down his arm and hers, over the dressings on the card table, toward the immaculate floor.

It was an inauspicious start to say the least; the valve meant to stop the blood when she pulled the tube away seemed to be faulty. They replaced the faulty device with another from their boxes, and from then on things went smoothly. Lots of people joked, but nobody fainted.

The team from Rochester had come to Michigan hoping to see at least fifteen new members of Mildred's family, and to send back as many blood samples by overnight express. After four days, they had sent ahead nearly twice their quota: thirty new people seen, and twenty-nine sets of blood tubes sent to Yale, to be analyzed in the laboratory of geneticist Kenneth Kidd.

The one missing set of blood tubes was marked for Mildred's nephew Peter, who was dark-haired, shy, good-natured, and all of eight now. He was the very last person asked to give blood, and he balked. Ernest came into the family room afterward, visibly upset.

''How'd it go?'' Como asked. She shook her head and slumped down on the sofa, her eyebrows drawn down.

Peter was charming and vulnerable. Tics rippled across his midsection. Peter took great care doing the pencil-and-paper tests. The school psychologist had diagnosed attention-deficit disorder. Como wondered if he was simply so busy ignoring his tics that tests always took longer than they should.

Laurie Ernest had enjoyed videotaping Peter. They'd talked about camping, about how much he loved to fish. Peter told her a funny story about fishing, and they both laughed. It didn't hurt too much to kill a fish, he added gravely, but he thought he'd never be able to hunt. Ernest's heart skipped at that, at the tenderness of a small boy. It made her long for her own small boys, nearly his age, who were back in Rochester with their grandmother.

After the videotaping, Como asked Peter if he would be giving blood. The boy did not respond, but his lip quivered.

Without waiting to hear any more, Laurie Ernest went upstairs to talk to Kurlan.

"I hate to coerce him," Kurlan told her. "But he's in such a perfect position in the pedigree to answer a question. His mom is Clare. That means every one of Mildred's siblings has the gene. Now Peter's got it. Which makes it pretty clear that Mildred's mom must have it too, as well as her father." He asked her to see whether she could gain his confidence.

She went back downstairs. Peter was obviously trying to control his tears. "Can I talk to you for a minute?" she asked, and he nodded. "In a short time, while you and I were talking, we had a good talking relationship. So I'm not going to play games with you now. Listen. Do you know what it means to be a scientist?" Peter's eyes brightened a little. "Well, you can be a scientist a little this afternoon, except it might hurt, just for a second, just like a little stick or a pinch. Could you do that?"

Large brown eyes looked back at her, and again she saw her own boys in them. "Could you do it?" he asked her. "I'd be delighted to do it," she replied.

He was very quiet. "Could you or my father hold me?" his voice quivered.

"I can't hold you and draw blood. But your father could surely hold you."

"Do I have to do it?"

"No, but it would be a great help if you could."

Peter was silent for a moment. "Can I be alone with my father?"

"Sure," Ernest said, and left the room to find him. As soon as the door closed behind her, she heard a hiccup, and then a small, deep wave of sobs. She hurried up to the family room and motioned to Peter's father to go downstairs.

"Roger," she said, "I think we crossed the line here. Even if I have him sit down and give blood he's going to be sobbing, and I don't think we want that."

Kurlan went down the stairs, opened the door to the back room, and said, "Blood drawing time closed. It's candy time."

"Early Christmas," said Ernest. She gave Peter a handful of jawbreakers, then a kiss and a hug.

■   ■   ■   ■

A few minutes later, the team began to close up shop, to return their materials to the cartons. They filled a trash bag with discarded paper and medical supplies, and returned the Lego sculptures to the middle of the billiard table. Mildred's son sat nearby, looking a bit forlorn. The occasion had made him feel important, Mildred had told Kurlan; now it was over, and his problem was just a problem again.

After saying farewell and thanks to the van Dykes, the three researchers got into the station wagon and drove toward Lake Michigan. They wanted to have dinner at a seafood restaurant, to relax, to celebrate, to take stock.

"You want to go take a look at the dunes?" asked Kurlan, as they approached the restaurant near the waterfront.

"Sure," said Ernest from the backseat. "As long as I don't have to get out of the car." They continued toward the lake shore. It was an ominous autumn afternoon. The sun had set; a deep bank of black clouds was moving eastward, and leaves were clattering across the streets.

The drive wound under a canopy of yellow leaves. At length it opened into a parking lot. "There they are," said Kurlan. "There are the dunes." To the right, covered with very small trees and scrub, the sandy hillocks rolled toward the land. To the left, beyond a double length of snow fence, gray breakers crashed onto the beach. The clouds were closing in.

"And there's Lake Michigan," he added.

Even on a clear day it might be an ocean; you cannot see the other side. That evening, the clouds were low over the lake. The three people sat for a moment in the car, looking out over wet sand, across the rolling whitecaps, toward the point where gray-green water disappeared into mist.

They were all pleased with their success. By the next day, all twenty-nine samples had reached Ken Kidd's lab in New Haven. The rest was up to Kidd and Pauls. They might find a linkage the day after that, the next year, or not at all.

If a linkage did emerge, it could vindicate David Pauls's limited definition of Tourette's syndrome. Of course it might instead substantiate the Comings' view, that a very common gene exists that predisposes a person to anything from tics to drug abuse. What would happen then? We shall see.

5

THE EDGE OF DISCOVERY

To follow the legacy of disease in the blood, scientists must extract the DNA, break it into fragments, and see whether a certain kind of DNA fragment coincides very often with the presence of the disease.

Many important questions intervene. How often is "very often"? How should you define the disease in the first place? What does it mean if some families show the pattern and some don't?

One well-known study among the Amish has crystallized some of these questions. In the process, Amish DNA pinned a mental illness to the gene map, and later took it away again. Maybe.

Such are the caprices of modern research in human genetics. It's exciting, it's full of potential, and it's very risky.

In the autumn of 1959, when a fledgling sociologist named Janice Egeland began fieldwork among the Old Order Amish of Pennsylvania, her first question was where to live. Several Amish families offered to rent rooms to her, but she declined.

"Some voice told me that any person who would take someone in for money cannot be in totally good standing," she explains. It was a sound instinct: One of those people was later excommunicated. She rented a room from a non-Amish farmer, and began her work.

She would loiter at places where the Lancaster County Amish congregated, such as horse auctions and dry-goods stores, hoping they would begin to see her as a neighbor. But the Amish are hardly gregarious with strangers, and this approach was not a success.

Egeland had a fellowship in a new program at Yale University: medical sociology, the study of health care as an aspect of human society. She had hoped to become a doctor, but didn't have the money. This was an attractive substitute. She came to Lancaster County, hoping to study in one of America's last surviving clans the factors that prompt some people to choose ''alternative medicine'' while others use ordinary doctors.

In search of contacts, she tried to find members of the Old Order Amish who had ever worked at local hospitals. One autumn afternoon, given a tip about such a person, she stepped onto the front yard of an Amish farmhouse, briefcase in hand.

A woman in a black apron and bonnet stopped pushing the hand lawn mower long enough to call out, ''I'm sorry. We don't want any magazines.''

Egeland would have smiled. ''I'm not selling magazines,'' she replied. ''I've heard that someone in your family once worked at the medical center. I'm doing a medical study. I thought perhaps you might be able to help.''

Worldly magazines are one matter to the Amish; medical studies are quite another. The young stranger with the oval face and spectacles looked very earnest. The next thing Egeland knew she was in the large kitchen, chatting over sticky buns and coffee.

They talked for a long time. Without telephones and television, many Amish people see conversation as a major amusement. Some members of the woman's family, like many Amish women in particular, enjoyed talking about ailments. When Egeland walked back across the half-mown lawn, it was dark.

In time, Egeland came to call this woman's kin her ''special family.'' Of all the Amish families she later came to know, they would mean the most to her.

They sold goods at the farmer's market, which made them

more comfortable with outsiders than some members of the community were. A keen interest in health made them kindred spirits to Egeland. In time, they invited her to move in for free. But what made them most special came out much later, and had nothing whatever to do with her Ph.D. thesis.

Egeland's special family—and one person in particular, who became a dear friend—were marked with a severe psychiatric disorder. Probably the last thing she would ever do is to give particulars about her special family, but we do know something from their pedigree, which she has identified from time to time.

They are descendants of a man who died in 1898, leaving 8 children, 5 of whom had psychiatric illnesses. They gave him a total of 35 grandchildren, among whom 8 were mentally ill, and 191 great-grandchildren, 25 of whom had psychiatric illnesses. Egeland has traced the family back to a key progenitor born in 1763 who, according to written documents, had depression. His brother committed suicide.

Egeland says that the Old Order Amish regard manic-depressive illness as the worst disease humans can suffer, worse even than cancer, "because it interferes with the function of the mind and the wholeness of the body." The more she came to know her special family, the more deeply she came to share that conviction.

Ultimately, she staked her entire career on the study of that disease, which she prefers to call bipolar disorder, reflecting its rapid swings from euphoria to desolation. She spent decades tracing its hereditary path through some Old Order Amish families. For a nonpsychiatrist, it was quite a bold venture.

The Amish people began to emigrate to North America in the early eighteenth century, in search of religious freedom. They had descended from a group of Swiss Protestants, the Anabaptists, and traced their religious roots to the sixteenth century, the first days of the Reformation. Their fundamental doctrines, adult baptism and pacifism, were radical enough four centuries ago to provoke exile or execution, sometimes by burning at the stake. Many Anabaptists who survived fled to America.

The Old Order Amish are the most conservative of dozens of sects (including the Mennonites) that descend from the Swiss

Anabaptists. There are no Amish left in Europe today; their oldest and largest settlement is Lancaster County, Pennsylvania. Today there are some 12,500 Old Order Amish in the county.

The town at the center of their community, Intercourse, was founded in 1754. Its remarkable name probably derives from an ancient racetrack, once located between two stagecoach stops. This is a little-known detail about the name Intercourse, not nearly as amusing as its worldly contrast with the images usually evoked by the word *Amish*.

In 1954, the merchants of Intercourse sponsored a bicentennial celebration. They of course provided souvenirs, among them buttons and bumper stickers with the legend: "Intercourse, Pennsylvania—Amish Country." The buttons and decals were big sellers, and must have traveled all over the United States. One Amish historian places this as the start of the county's slow transformation, from a placid, rural backwater to an unofficial theme park overrun by tourists and featuring the Amish, to their complete dismay.

Janice Egeland arrived five years later, in 1959, to begin her work among the Old Order Amish. At the time, Intercourse was hardly more than a mile-long main street where people from surrounding farms came to shop. But they were already beginning to lose their blessedly isolated Utopia.

These days, for most of the year, the roads around Intercourse and Bird-in-Hand are jammed with out-of-state cars. There are "Amish" motels, "Amish" all-you-can-eat restaurants, guided tours through "real" Amish houses and farms. Twenty minutes away, a strip of discount factory outlets has sprung up along Route 30.

When Janice Egeland arrived before 1960, however, Amish Country did not yet have the name. It belonged to the Amish, and they had far less reason to mistrust or dislike strangers. Nobody was aiming a camera at the haying wagon; fewer cars were passing the buggy on the upgrade.

It was a very special time, a peaceful period in her life. Egeland grew to know the plain-spoken humility and kindness the Amish are famous for. It springs from what their European ancestors called *Gelassenheit,* or "yieldedness." One scholar describes it as "the power of powerlessness."

The Amish believe in rejecting the natural impulse to gain

what you want by using power; the goal is to do whatever is best for others. The same ideas are preached in other religions, but for the Amish the object is not personal holiness or a good feeling. It is brotherhood and community.

This explains why they mistrust technology. Cars and electricity aren't sinful, a bearded Amish historian told me with a twinkle in his eye. It's what they accomplish. When errands and chores are too easy, people stop depending on each other. That threatens the community.

Although the Amish don't evangelize, their way of life can be seductive. To Egeland, it was irresistible. As a sociologist, she admits today, she began to lose it, to abandon her detachment and sense of professional isolation. She says she fell in love, with an entire society.

"I got into the community," she once remarked, "to the point of risking my identity."

By 1961, Egeland was fairly well accepted as part of the Amish community and had become acquainted with another sociologist, John A. Hostetler, who was completing a study of Amish life. It was his book that drew the medical geneticist Victor McKusick into studying the Amish. McKusick got to know Egeland, and she began to help with his studies, ascertaining cases and coordinating fieldwork.

They were pioneering studies in human genetics, and were the foundation of McKusick's eminent career in genetics. He and Egeland parted company in about 1965, and Egeland completed her thesis. She had gained much from the collaboration with McKusick: She appears as coauthor on several notable genetic studies, and she also drew two important insights from him.

One of them appears in McKusick's classic book about his studies of the Amish, and is often quoted. "I go to them as a physician, not as a geneticist," McKusick stated, "and . . . as a physician genuinely concerned about those among them who are physically and mentally handicapped."

However genuine, it is a very useful strategy. Not all of the Amish are entirely taken in by it; they know that researchers can gain status and also grant money from the cooperation of the Amish. But like anyone else who is helpful at heart, they do appreciate getting something back for their effort—in this case, medical care.

Besides providing the research strategy of quid pro quo, McKusick impressed upon Egeland the way in which the Amish could be most useful to science. "I remember him saying, this is a genetic gold mine," she said, years later. "That went down very deep in a young graduate student looking for a niche to fill."

More immediately, she found at least a temporary niche: the need to update the Fisher book. This complete genealogy of all the Amish of Lancaster County, which had appeared in 1957, was rapidly getting out of date. Although it is no different from any ordinary family tree (but for its size), the Fisher book seems to be held in higher regard by the Old Order Amish of Pennsylvania, as if without it they would not be who they are. They also enjoy using it to trace the tangled web of descent, as a sort of game.

To update the Fisher book, Egeland had to gather new information about the births, deaths, and marriages within the extended families. The best place to go would have been the church—except that the Old Order Amish have no church buildings, and their official register is the Fisher book itself.

Instead, Egeland sought help from one individual in each church district, usually a single woman like herself. She called them scribes.

She could not telephone her scribes; the only way to debrief them was to visit. It was coffee and sticky buns again, and the major Amish form of entertainment, conversation. The debriefing sessions went far beyond the usual vital statistics; the scribes shared stories about weddings, about a new baby or a long-awaited death, or perhaps about someone who was "nervous" or "too talkative." Janice Egeland, an eager and sympathetic listener, began to know the Amish nearly as well as they knew each other.

In 1972, the year in which the new version of the Fisher book was published, Egeland's father developed leukemia. He went for treatment to a nearby medical center, where he was transfused with over 900 pints of blood. He died shortly after Christmas. The bill from the blood bank came to $27,000—$30 a pint.

There was no way Egeland could repay it. Within days, Amish people began to arrive at the blood bank, in groups of ten, waiting to donate blood to erase the debt. More than five hundred Amish people eventually gave blood for Janice Egeland, and the heartwarming story made filler for newspapers from Miami to San Diego.

Afterward, Egeland took a leave of absence from her teaching job at a medical center in Hershey. She was exhausted and emotionally drained, and she needed time—not just to recuperate, but to prepare for a major change in her career.

She hardly knew how to proceed, but Egeland had decided to study mental illness among the Amish. Her father's death contributed to this decision: It had been a compelling counterpoint to what she had seen in some Amish families.

"When I saw my father's death, it was a terrible kind of dying, terrible hemorrhaging," she told me. "Still, his mind stayed clear. He was very much in control, of his feelings, of his communication with us as a family. He had a healthy outlook for what he could still contribute, even as an object of study. He could die with dignity."

All that time, she had an intimate knowledge of another reality, from her "special family." It had been her first close contact with serious mental illness. She was not a psychiatrist, and could not help.

All she could do was witness the ponderous bleak reality of bipolar disorder, the opposite of the qualities she loved about the Amish—or, perhaps, the vacuum left behind in their absence. She followed its swings between unpredictable brashness and deep gloom, and saw how a derangement of such an ineffable thing as mood could sap an entire family, the healthy as well as the affected.

"She was not a psychiatrist or even a Ph.D. psychologist," says Abram Hostetter, a psychiatrist who later collaborated closely with Egeland in her depression study. "To get involved with this was most shocking. To live with it, it sort of pulls you down."

As time went on, while Egeland completed the Fisher genealogy, she was struck by the way mental illness appeared to bedevil certain families in particular. "There are weak families and strong families," said one Amish farmer, who told her that several of his forebears had suffered severe mood swings. ''Mine is one of the weak families,'' he went on.

Egeland responded that some of the ''weak'' people with bipolar disorder were also some of the most productive. But he saw another side of it.

''That's as may be,'' he said later. ''But being productive is not so much, you know. The strong families may not be so productive, but you'd rather be around them.''

Until she saw bipolar disorder, Egeland says, she had been leaning toward a study of cancer in the community. ''But after living within a family setting and witnessing a person ill with this disorder,'' she said, ''I had no choice.''

A Bible verse and a short poem appear at the front of the 1988 version of the *Fisher Family History.* ''Though our outward man perish, yet the inward man is renewed day by day,'' reads the verse at the head of the genealogy, from 2 Corinthians. It is echoed by the poem that follows, which reads in part:

What though I falter in my walk?
What though my tongue refuse to talk?
I still can tread the narrow way.
I still can watch, and praise and pray.

An Old Order Amish person who has bipolar disorder cannot still ''tread the narrow way.'' The manic phase sometimes makes him excessively loud and brash; he may buy a forbidden tractor or a car, or argue with the bishop during a sermon. In the depressed state, he cannot bring himself to care about praising or praying, or about the work he must do to show he cares for his fellows.

Whatever it may say in 2 Corinthians, the ''inward man'' with bipolar disorder is not ''renewed day by day,'' unless he receives professional help. At its worst, as in those four families among the Old Order Amish of Pennsylvania who suffer the most severe consequences, the illness impels its victim to take the step that the Amish view as eternal divorce from God: suicide.

Bipolar disorder is classed as an ailment of the ''affect,'' or mood, as an *affective disorder*. It afflicts perhaps one in a hundred

North Americans. As Egeland is always careful to point out, bipolar disorder is no more common among the Old Order Amish than among any other group of people; it is not an "Amish disease."

Among the Amish, however, bipolar disorder is much more obvious than elsewhere. Amish people don't react to depression with drink or drugs. They won't have extramarital affairs or beat their wives. If an Old Order Amish man is depressed, he will probably weep and say he feels low. No one in his culture ever suggests he should not.

Abram Hostetter has treated many Amish patients in his private practice in Hershey, Pennsylvania, and has seen others as a consultant to Egeland's study. He says bipolar disorder arouses terrible guilt, often leading to confessions of imaginary sins, to a belief that the soul is lost, to nihilistic ideas, thoughts of doom, constant hand-wringing. In the worst of the depths, there is disinterest in any kind of activity. The victim becomes one who simply sits in the corner and stares down.

Of the other extreme, mania, Egeland has often given examples. They aren't the sort of behaviors that would stand out elsewhere, but among the Old Order Amish they are unmistakably deviant: running out barefoot to make a long-distance phone call, taking a vacation at planting time, bragging, flirting. To the Amish, "mania is not cute," she says. "It stands out starkly against what I otherwise call a bleak landscape."

During the leave of absence after her father's death, Egeland relocated to the University of Miami Medical Center, where she took a part-time job assisting in a study of health practices among blacks and Caribbean immigrants in Miami. She spent the remainder of her time preparing to write a grant proposal for a study of the genetics of psychopathology among the Amish.

In the early 1970s, the idea of genetic studies in psychiatry was outlandish to most people in both fields, genetics and behavior. Genetics could offer psychiatry little more than correlations, studies of whether a particular condition varied between populations or coincided with obvious traits such as age, gender, or alcoholism.

"Though we all had in mind that it would be great if we could find a single gene, the only way we could reasonably look toward that was by looking at the proportion of illnesses in different kinds of patients," says Elliott Gershon, the head of psychogenetic studies at the National Institute of Mental Health. "The first linkage marker study actually came out right then, in 1969. It was a linkage of color blindness to manic-depressive illness. But it didn't have the impact it should have had."

It did have an impact on Egeland, however. "Affective disorder in which mania occurs," wrote the authors of that study, in a book on bipolar disorder, "is quite probably linked on the X chromosome with the locus for color blindness." She hugged the book to herself, she says; it validated, in print, the idea she had in mind.

If the idea was unorthodox, it was especially so for someone who had credentials in neither field. So Egeland sought out commitments from people whose qualifications were impressive. One of them was Jean Endicott, a psychologist at the New York State Psychiatric Institute. Endicott had headed a project for the NIMH to create uniform standards of diagnosis for affective disorders, including bipolar disorder.

Endicott replied to Egeland's letter by phoning her from an auto repair shop on upper Broadway. A phone conversation with Janice Egeland is never brief; Endicott recalls they spoke for nearly an hour. Although Egeland had the wrong credentials, Endicott quickly realized that she was "exceptionally well versed" in psychiatry and study design, and that she had spent a lot of time reading and thinking to prepare for the study.

"She had done as much reading as most people with clinical experience," Endicott recalls. "She was completely familiar with the diagnostic criteria, as well versed as anyone I know. I was convinced that if anyone could do it, she could carry it off." In fact, Endicott realized, if Egeland didn't do it, no one else would.

Egeland spoke about her primary concern: the fact that she could not remain "blind," ignorant of her subjects' mental state, and therefore could never be sure to be objective in analyzing genetic tests of her Amish contacts. At the same time, Endicott says, because of the closed nature of the community, "There was no way to send in a cadre of other psychiatrists to independently interview her subjects. . . . So she had proposed having a board of diagnosticians without any information about the families themselves.'' Egeland intended to interview the subjects herself, and then send coded, anonymous descriptions of the interviews to several psychiatrists for their opinions.

"During that conversation, she asked if I would be willing to train them, and if I would be an outside diagnostician," Endicott recalls. "I said not only would I, but I would do it for all the cases, so that my diagnoses could be compared with the board's."

By that time the chairman of the psychiatry department at Miami, who had shown interest in her work for years, had made her an irresistible offer: If she could gain funds for her study, he said, she could be a visiting scholar. She could return to Pennsylvania, with no obligations to attend staff meetings or other functions at the University of Miami.

I have contacts among the Old Order Amish now, she argued to the NIMH. If we wait, I may move on to other commitments. I may lose my contacts. Both the Amish and the outside world stand to gain if I proceed as proposed.

However deep her convictions, they did not persuade the committee assigned to review the long proposal. Committee members felt that the study would be biased, board or no board, because Egeland knew her Amish subjects personally. Above all, the members—most of them eminent psychiatrists—felt her credentials were not adequate. She was neither a geneticist nor a psychiatrist. They scored the application too low to receive funds.

Normally employees of the National Institutes of Health are not directly involved in deciding who should get grant funds. But on rare occasions, they may overrule the review committee. Egeland's convictions did sway Benjamin Locke, the chief of epidemiology and psychopathology at the NIMH.

"Despite those two problems, I made the judgment that there are only a few populations suitable for this kind of research," he says. "The study had a whole host of positive aspects." His every impression was that Egeland was extremely conscientious and capable; she was doing everything she could to avoid misusing the data or directing the results of the study.

"Despite the concerns of the review group," he says, "I decided to stick my neck out. I got permission to fund this."

In September 1976, the study began.

■   ■   ■   ■

That autumn Egeland sought out her scribes again, with a new agenda. In part, she intended to update the genealogy. This was a familiar role, and one that would bring no comment. But as she met with each of the scribes in private, Egeland said she would be seeking additional information, beyond the census data.

She began to ask about the mental status of each individual, in search of probands, affected individuals from whom to trace lineages of mental illness. She asked the scribes to repeat anything unusual they heard from day to day that might be useful to the study. Over time, she began to hear the same reports about the same people from several different sources. She supplemented these reports with records from local doctors and hospitals.

Slowly, Egeland began to seek introductions to the mentally disturbed. Informal family visits were as useful to her as direct interviews, and much easier to arrange. Because many Amish people enjoy talking about health she could expect useful details to emerge of themselves. Gradually Egeland began to seek out close relatives of the potential probands, and to ask what they knew of the behavior of others in the family who were no longer alive.

Such information never comes to someone who knocks on the door, clipboard in hand, and starts asking questions—least of all in Amish country. Egeland knew that it had to be a long process. She could never phone ahead; she had to drive by to see if the family was in. If she needed to see the father, she probably had to arrive near dawn, before milking time.

"She had a way of just getting around to it," says one farm wife whose family participated in the study. "She would start chatting way out here"—and she reaches her hand into the air—"and slowly get around to it, until you didn't even think about it."

By all accounts, she was a wonderful listener. To many people, obviously, she became a friend. In time, she became the person to contact for help, if someone began to get out of control and started driving the horse too hard or staying up all night to clean house.

Within four years, Egeland discovered 102 people among

the Old Order Amish—57 women and 55 men—who showed signs of bipolar disorder. Looking back, she thinks her thesis research made this easier. The study had shown that what people believe about their health has little impact on what they do about it. What matters is the way other people they respect behave when they are ill.

"Someone else in my role would become offended because people will say this mental illness is due to low blood sugar or because they had their hair analyzed and it has too much aluminum," she told me. "I think if I didn't have this broader view, I would have lost time trying to overcome these ideas, which is so frustrating."

And unnecessary, she has found. "I just try to get these people to a doctor," she went on. "For them to get good results is enough to keep them returning and taking their medicine."

From the outset of the study, there was good reason to believe that the community would be supportive. Amish leaders were deeply disturbed by the few suicides that had occurred, and especially by the alarming proportion that had claimed church leaders. The prevalence of suicide was well below the national average. But the fact that it existed at all impelled the Amish to seek an explanation.

Among the Amish, Egeland reported in the *Journal of the American Medical Association,* suicide is starkly obvious. The cause of death is never as ambiguous as a drug overdose or a car accident. A desperate determination is unmistakable in the details of the suicides among Old Order Amish in the last century.

There were twenty hangings, usually in the barn, where they were unlikely to be interrupted. All gunshots were to the head. One man drowned himself by tying a stone to his body and jumping into a deep pool. A woman was found floating in the washhouse cistern, which she had reached by climbing up and over a boiler, Egeland wrote, "making accidental access impossible."

If there were no other result of the study, Egeland often said, she hoped it would lead people to regard mental illnesses as biochemical disorders, not as personal failings. For the Amish community, it apparently had another outcome as well: Since the study began in 1976, there have been no suicides among the Old Order Amish of Lancaster County.

■   ■   ■   ■

Slowly, in batches of ten, typed reports of the interviews began to arrive in the consultants' mailboxes. They were correct in psychiatric detail but misleading in other ways. Families were jumbled, and people were identified with anonymous titles, descriptions like "potato digger" or "woman with red socks" rather than names.

Egeland sent them in no particular order, normal cases always mixed in with the affected ones. After the cases became too numerous to recall, she began to cross-check her panel's reliability by sending some previous interview reports again, under a different description. Periodically, the panel would meet with Egeland and Endicott to clarify the confusing cases and arrive at a consensus.

Bipolar disorder was obviously distinctive in the Amish, but was it really hereditary? After amassing details about some thirty-two families with a high prevalence of psychopathology, Egeland sat down with another set of consultants to decide which family had the most useful structure for genetic studies in search of a linkage. They were looking for a large extended family, with numerous siblings among whom the illness was common, in which everyone from children to grandparents were accessible and willing to cooperate.

In the end, Egeland's "special family" did not fit these criteria best. The most useful family structure was what she had identified as kindred 110, with more than 80 family members, 19 of them affected by bipolar disorder.

Even before any one of these people gave up a vial of blood, it was essential to judge from the pedigrees whether bipolar disorder really followed a predictable path of descent. The person who completed this analysis, like all of Egeland's collaborators, came from a list of specialists recommended by her committee of advisers at the National Institute of Mental Health. The molecular geneticist she chose from among them was Kenneth Kidd, at Yale. The statistical and analytical work on the pedigrees became the responsibility of his co-worker, David Pauls. (For both Pauls and Kidd, the studies among the Amish preceded their research on Tourette's syndrome among Mennonite and Dutch-American families.)

Pauls's first task was to carry out a series of *segregation anal-*

yses on the Amish families. These are a marriage of genetics and statistics, a method worked out during the last several decades. The process has been transformed by computers, which can store large sets of data (such as family trees) and evaluate them in several different ways in a very short time. A number of computer programs exist for segregation analysis; they all assess whether any of several proposed genetic patterns actually explains the inheritance of a trait in a particular pedigree, beginning with the rules of Gregor Mendel and taking in many other factors.

To do the analysis, Pauls had to begin by making certain assumptions from prior knowledge about bipolar disorder. The most important is that a gene or genes is in some fashion responsible. But what is the gene's *penetrance:* how often do the symptoms emerge in a person who has the gene? How common is the gene? The analysis depends on best guesses about all of this.

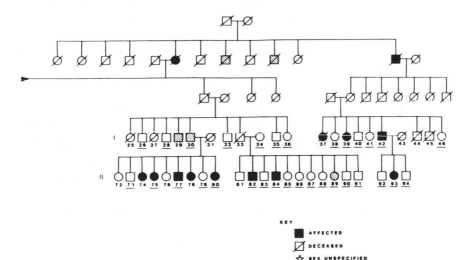

Part of the pedigree of a large Old Order Amish family with a high proportion of bipolar disorder. The complete pedigree includes 301 individuals. (From the Coriell Institute for Medical Research, where the cells of this line are preserved; with permission)

For something like eye color, gene penetrance is 100 percent, but for many other conditions, such as cancer and heart disease, it may be lower, depending in part on factors such as diet and exposure to chemicals. Pauls also varied his analysis by including different subtypes of mood disorders in different models.

He began by comparing the two simplest explanations for the pattern of bipolar disorder: that it is not inherited, or that it is inherited in a dominant fashion, so that anyone who possessed a single copy of a single gene would be susceptible to the disease. The statistics showed that the second explanation was far more likely than the first, by odds of one to a million.

In fact, that pattern of transmission was more consistent with the occurrence of bipolar disorder in kindred 110 than was any other pattern. The analysis also helped to predict which other psychiatric conditions related to depression might be caused by such a gene, and which were unrelated.

With her pedigree chosen and her segregation analysis complete, Egeland returned to the National Institute of Mental Health with a progress report and a request for more funds. She proposed to test the families for a few traits that had been linked to the heredity of bipolar disorder. One was color blindness; another was a blood type, Xg^a.

Yet again, the review committee gave her a low score, for much the same reasons as before. Ben Locke paused to reconsider. He went to Elliott Gershon, who said, "By all means. Continue funding her." Again, Locke requested special permission to continue funding the study, and again it was granted.

Elliott Gershon helped in another respect, which was going to prove just as critical: he had put her onto a good home for the Amish blood samples. This would have been irrelevant to the Amish; once the blood was spared, in general, they forgot about it. They would have regarded it in the same way they regard farmland they have sold. If it is out of their hands, it is in God's hands. Let God help whoever uses it, they say. Whoever misuses it, God will take care of him, too.

Janice Egeland could hardly afford to be so disinterested. As she loves to point out, she straddles two fields: the agrarian and the academic, the cornfields of the Amish and the research field of molecular biology. In one field, the philosophy of life is yieldedness, or *Gelassenheit*. In the other, whatever its workers may say for the record, the philosophy has been quite different. When it comes to yielding up items such as cell lines and sharing results, the community of molecular biologists would not bring Amish farmers to mind.

As to the competitive and secretive habits of researchers in molecular genetics, Janice Egeland was then as innocent as her Amish subjects. Gershon, at the center of psychiatric genetics, was far more savvy—and also, incidentally, interested as an outsider in seeing to it that no one gained exclusive control over the invaluable resource Egeland was about to create.

The University of Miami Medical Center had no facility for making archives of blood cells at the time—and if it had, Egeland told me, it might have wanted to be as protective as anyone else, to save the resource for its own scientists and their collaborators. Ken Kidd, the designated molecular geneticist on the study, did not have a place for them, either. He had taken a sabbatical in a laboratory in Boston, expressly to learn techniques such as those needed to build up a cell line, but when the Amish blood cells were ready, he was not ready to receive them.

Meanwhile, a national cell repository in New Jersey had approached Gershon, seeking to amass a collection of blood cells from a large, informative family with a psychiatric disorder. "This would be a nice opportunity for you," Egeland recalls Gershon telling her. Instead of having to choose which labs would receive her samples, he pointed out, she would have the privilege of letting someone else preserve and distribute them. Her own co-workers would have a chance to study them exclusively for a year; afterward, anyone would be able to buy the cell lines for his own research.

Until Gershon mentioned it, Egeland had never heard of the cell repository in Camden, New Jersey, at what is now the Coriell Institute for Medical Research, an independent laboratory and resource for medical scientists. The repository has pioneered methods of immortalizing blood cells in tissue culture, so they can be kept in cell culture indefinitely.

Quickly, she made contact, and her offer of the Amish cell lines was accepted. Today, Egeland regards the presence of "her" cell lines in the Coriell collection as something like a permanent grant, because of the considerable expense required to establish and store any single cell line.

In addition, it proved to be something of an insurance policy. "I was embarking on moving into the molecular level, the DNA level," she says. "I think if any one lab had gotten a hold on such a set of materials, they would have wanted to be very possessive. If I'd had a lab, I would have felt that way too. As it was, I could step beyond providing materials for any one lab, and running the risk that any one lab might be too possessive. I had a chance to facilitate more and perhaps better scientific work, competitively."

Between 1979 and 1982, Kidd and others tested the Amish blood samples against 42 blood types and 17 immune system markers, looking for a pattern that matched the inheritance of bipolar disorder. All of the results were negative.

"The absence of a positive finding in this study is not sur-

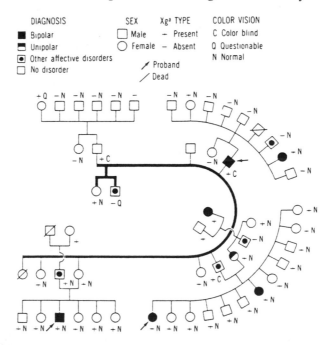

Various protein markers not linked to bipolar disorder in an Amish family. (From Kenneth Kidd, Janice Egeland et al., "Amish Study IV: Genetic Linkage Study of Pedigrees of Bipolar Probands," *American Journal of Psychiatry* 141:9, September 1984, p. 1043; with permission)

prising, given the paucity of genetic markers used," Kidd and Egeland wrote when they published their results in the *American Journal of Psychiatry*. When the study began in 1976, they pointed out, there were only about 30 "reasonably informative" markers, or known polymorphisms, among the 23 pairs of human chromosomes, all of them blood or protein markers. The odds that "the presumptive locus for bipolar affective disorder" would be close to any one of them were very low.

"The situation has changed dramatically as part of the revolution in human genetics," the authors went on. "The rationale for an extensive linkage study is clearer and more compelling than ever."

In late 1981, Egeland gave a seminar at the New York State Psychiatric Institute for Jean Endicott and her colleagues. She presented her data and expounded upon affective disorders among the Amish. As always in public, she was warm and upbeat.

The human genome should be saturated with new markers within a decade, Egeland predicted, so finding a linkage for bipolar disorder was only a matter of time. She was astute: two years later, by 1983, there were 200 known genetic markers (variants of DNA itself, as opposed to traits such as color blindness, or blood types such as Xg^a). The genetic markers were landmarks along the human chromosomes, detectable by treating DNA with restriction enzymes. By 1985 there would be about a thousand of these landmarks, which could lead researchers to the locations of important genes—including those for mental disorders.

"It's a whole new view of what psychiatry can and will be," she said. If they could only begin to find the genes behind mental disorders, the identities of the faulty brain proteins could soon be unmasked. An amazing prospect loomed beyond that, of using blood cells to diagnose or even predict susceptibility to mental disorders. It was not an entirely new dream, but for the first time it really seemed to lie in a future within the limits of our vision.

"We're closing in, and that's the exciting thing for everyone," she said to me at the time. "It's possible to be at the edge of discovery on this sort of thing."

■   ■   ■   ■

For a year, the search for new DNA markers in the blood cells from the Amish was a challenge that belonged to Kenneth Kidd alone. An intense, bearded man of forty, with all the compact energy of an elf, Kidd was well placed to do the study. As a tenured associate professor at Yale, he had a permanent position in a prominent place. Another sign of his stature was his appointment to a the mammalian genetics study section of the National Institutes of Health, where he was responsible for reviewing grant applications in his own field.

As soon as he learned of the first DNA linkage to the sickle-cell gene, Kidd says, he was certain that other such polymorphisms would be found in DNA, and that they would be linked to genes at the root of other diseases. He was convinced that they would also lead to new insights into our brains, and reveal some order behind the baffling array of abnormal mental conditions.

By the late 1970s, he had published genetic studies related to schizophrenia and stuttering. He hoped to find a DNA landmark for the location of a gene behind mental illness, and the importance of the Amish blood was obvious to him. But Kidd was well aware that he did not know enough molecular biology to take full advantage of the Amish cells. To test them for new genetic markers, he would have to retrain.

So he went on sabbatical, for a full academic year, as a visiting scientist at the Massachusetts Institute of Technology, in the laboratory of geneticist David Housman. Housman was also on the NIH study section, and shared Kidd's enthusiasm about the future of human gene mapping. He also knew the latest methods of molecular genetics.

Kidd proposed to share some of his own DNA with Housman and his co-workers, in exchange for training in the molecular techniques he needed to master. Housman's own primary interest was the genetic control of hemoglobin production. This focused his attention on chromosome 11, where the hemoglobin gene was located.

For his own part, Housman was intrigued by the new possibility of finding a linkage to a psychiatric disease. "Affective disorder would be a real bonus if we saw an association," he said later. "But we didn't expect it to happen real quickly."

At this point, the nature of the study had changed, and not

subtly. Janice Egeland had begun her work because she cared deeply, and personally, about the Amish. If nothing else, her continued success among them demanded that she be seen as caring deeply, as she obviously did.

But her collaborators at Yale and MIT were distant in every sense. As Housman puts it, helping people who are ill is "not the fundamental drive" for most laboratory scientists. They want to solve puzzles.

Kidd left the Housman lab in the late spring of 1982, and returned to Yale, taking the Amish cell lines with him. But he left Housman's lab with some samples of extracted DNA, and an agreement. Since the Housman lab was already interested in chromosome 11, they would focus on that one chromosome alone, testing markers on that chromosome for a linkage to bipolar disorder among the Amish. A new postdoctoral fellow in the lab, Daniela Gerhard, had taken up the task. Meanwhile, as the people in Kidd's lab refined their own ability to analyze DNA, they were to begin the arduous process of matching the Amish DNA against known genetic markers on all the other chromosomes.

As Kidd and Gerhard worked, they waited for a pattern to emerge. They relied on Pauls's computer studies to assess the likelihood—ten to one, a thousand to one, a million to one—that any pattern that emerged might be a chance occurrence, a fluke.

Testing blood for genetic markers is a tricky task, a multistep process that matches DNA segments in a logical sequence of finicky operations. It involves two basic procedures: (1) extracting DNA from the blood cells given by a family, and breaking it up with enzymes so that it is usable; and (2) exposing that DNA extracted from the family to another DNA sample, one containing many copies of one small genetic fragment that has been made radioactive. The small, identical radioactive fragments are known as *probes*.

In a liquid solution at the right temperature, DNA will always search doggedly for its mate, a reciprocal or complementary sequence of subunits. A probe will anneal to its own chemical mirror image in a geneticist's test sample of DNA from human cells, according to the invariant rules of DNA subunit matching—A with T and C with G. Because it is radioactive, the probe will darken a complementary sequence if the two are mated and developed on an X-ray film.

The probe is used to detect a mutation—sometimes a massive deletion of DNA, often a mere difference of one single DNA nucleotide. Such a mutation would create a genetic variation or polymorphism, a region of DNA that must occur in the human population in at least two forms (one containing the original DNA sequence and the other containing the erroneous one). About one of every thousand nucleotides is the site of such a variation. An error may make no difference whatever to the individual (especially if it occurs in a part of the DNA that is never "read" into a protein) but the scientist may use the variation to locate an important gene.

If a mutation has occurred at the spot where an enzyme breaks DNA, the enzyme will create different-size fragments of DNA from people who have the mutation than it does from normal DNA. The size difference can be detected on X-ray films because of the radioactive probe. If such a difference in fragment size—such a polymorphism—corresponds with the occurrence of the disease in a pedigree, the probe represents a new genetic marker for the disease. (See the illustration on page 74.)

The object is to test as many probes as necessary until you find one that shows a pattern that coincides with the occurrence of the disease you are studying. If a certain mutation shows up often enough in people with the disease, and seldom enough in their healthy relatives, it's a sign that the probe's sequence and the gene behind the disease are very close together on the same chromosome.

But how often is often enough to establish that a gene for a condition such as bipolar disorder is actually located close to the DNA fragment represented by your probe? And how close is close?

It's a matter of odds. The more closely a marker and a gene are linked, the theory goes, the less likely that they will be separated by accident as the DNA assorts itself into sperm or egg. They will tend to descend together, generation after generation; there will be a *genetic linkage* between them.

Because it's easier to compare single digits than large numbers, geneticists generally agree to speak in what they call *LOD scores*—log odds, or powers of ten. If the odds are one in a million that a marker and a gene are not actually linked on a chromosome (that what you have found is misleading), then the LOD score will be 6: 10^6 to one. If the odds that you have not really found a genetic marker are only 1 in 100, the LOD score will be 2: 100 to 1.

There's been a long tradition in medical research that odds of one in a thousand are good enough to rely on. So in genetic linkage studies, any LOD score above three is generally assumed to be something worth making noise about. The higher the LOD score, the louder the noise.

In late 1983, James Gusella's lab at Massachusetts General Hospital in Boston sent up a Roman candle: they reported in the eminent journal *Nature* that a marker on chromosome 4 coin-

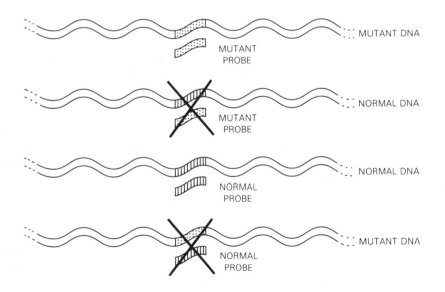

How probes spot mutations. The radioactive version of the mutant DNA will bind to mutant but not normal DNA, and vice versa.

cided reliably with the occurrence of Huntington's disease in that kindred of Venezuelans mentioned on page 90. The LOD score was over 7.

This discovery did not mean they had found the gene, but they had cornered it. Now everyone could begin trying to close in on the gene itself and, with any luck, on the cause of the incurable disease. Right away, they could test whether the same marker was valid in other families with Huntington's (which it was), and it could be used as a test to diagnose or even predict the disease.

For many months, Gerhard, in the Housman lab, had been turning up negative linkage results for bipolar disorder on chromosome 11. This was to be expected: With 22 other chromosome pairs to test, the linkage was most likely not to be on the one Gerhard had been assigned. Then, in early 1984, the computer spewed out something intriguing: a LOD score of 1.7, for a linkage on chromosome 11.

"Though as yet not definitive, these provocative results suggest that a major locus for affective disorder may be located on the short arm of chromosome 11, near the insulin locus," reads a report of the tentative linkage, "with odds of 50 to 1." The brief report was signed by everyone on the team, and the title was a question: "Is a Gene for Affective Disorder Located on the Short Arm of Chromosome 11?"

At the time, Kidd's answer would have been negative, even though he was the one who presented the report at a scientific meeting. "We presented it as not proof at all," he recalls, "but as tantalizing, positive LOD scores requiring additional work." As Kidd saw it, the report was worth presenting to confront people with the novel idea, soon after the Huntington's success, that the same techniques could be used to study genetics in psychiatry. But he doubted strongly that the bipolar disorder gene would show up in the first place they looked for it.

Even Gerhard herself was dubious. A science reporter came to see her shortly after the 1.7 LOD score appeared. "I remember telling her that even with 1.7, I'm not really thinking it's there," Gerhard recalls. "But I think maybe it's my responsibility now to do something about it."

As soon as she learned about the 1.7 score, Egeland too felt a responsibility to do something about it. She informed the authorities at the cell repository in Camden that the next set of cell lines should be split between Yale and MIT, 50–50. "I decided I would split the cell lines so that they had to talk together, cooperate together, act as a team," she said, "rather than one trying to undermine the other."

Long since, Gerhard and Kidd had been disputing about the disposition of the cell lines. Gerhard complained that Kidd was too slow to send her more DNA when her supply ran out. Kidd felt that the Housman lab was using the DNA much too quickly and perhaps too casually, considering all the effort that had gone into it.

Egeland heard about all this in Amish country, that place where a handshake is a commitment and where Amish farmers will spend all day raising a barn for a total stranger after a fire. The cells were, after all, the genetic essence of people she knew personally. They may have given up all right to the cells, but Egeland felt a powerful urge to ensure their proper use.

In Lancaster County, the Amish farmers were running out of farmland to pass on to all their children. The researchers who were studying some of their DNA, Egeland felt, should also be aware that they were farming in very limited territory.

"Here we have one large extended family," she remarked later. "Period. It makes you very humble. When you're in the hospital and see the devastation, you know it's one awful illness. It makes you think there ought to be an international panel of some sort: Gershon, Gerhard, the Yale people, God knows who else. . . . I would like to have the materials available to everybody. Everybody. If we care about the illness, people who can generate materials should put them out to the whole community, so they're accessible to everybody. Why should one lab rather than another be favored?"

Meanwhile, Egeland was taking her role as principal investigator most seriously. She was just as particular about the accuracy of the DNA testing as she had been about the psychiatric diagnosis. She wanted to see the results of the DNA studies—not just the numbers, but the actual laboratory films—and to have explanations of the meaning behind all the little splotches of radioactive signal. All this didn't make her entirely popular with her co-workers in the laboratory.

The field of molecular genetics is full of such frictions. It is complex and tedious, and large collaborations are unavoidable. To Jean Endicott, who witnessed many disputes firsthand, they were only to be expected.

"In collaborative work over a long period of time," she says, "you get to know a lot more about a lot of people than you otherwise would. Everything tends to go quite smoothly until there's something to publish, at which point a lot of variables come to the fore. That certainly happened in this case."

People had to be forced to share. But like so many efforts of ordinary humans working cooperatively, after a fashion, it proceeded toward a goal.

While the next thirty blood samples were being tested, Ken Kidd—still fairly certain that the tenuous linkage on chromosome 11 was likely to fall apart—went on vacation and then to a brief visiting appointment at a university in Australia. Back in Boston, the work he had no expectations for was rapidly amounting to something.

The additional samples from kindred 110 were quickly driving the LOD score upward, not down. The linkage was being confirmed.

At some point during this stage, Janice Egeland called the Housman lab, very upset, almost in tears. A woman in kindred 110, now a dear friend of hers, had suddenly and for the first time gone into a severe manic episode. Egeland was distraught for two reasons: first, that another member of a tormented family was afflicted, and second, that (she assumed) the change in her status would drive the promising LOD scores down.

On the other end of the line, she was stunned to hear that the news she had was good. Still totally blind to the DNA scores of the members of her kindred, she could not have known that her friend had been one of the few who fit the linkage pattern but did not show the symptoms.

The LOD score now approached 4, placing the gene for bipolar disorder somewhere between the location of the insulin gene and that of HRAS1, a gene that had been linked to the development of cancer. Near that gene, the LOD score was close to 5, odds of 100,000 to one that the first genetic marker for a psychiatric illness had been found.

Egeland, Pauls, and the scientists in Housman's lab began to write up the results for submission to *Nature*. Isolated in Australia, Ken Kidd received copies of the manuscript by Express Mail and electronic mail. He had gone to MIT in the first place to learn how to find that linkage; now the MIT team had found it, and not he.

The report of a genetic linkage between bipolar disorder and a site on the short arm of chromosome 11 appeared in *Nature* on February 26, 1987. "We see this as a landmark study," David Pauls was quoted as saying, in a companion article in the news section of the journal. Words within the article itself were equally upbeat.

"It is clear that DNA-based diagnostic procedures have a future in psychiatry," wrote the authors, a list of eight names that began with Egeland, proceeded to Gerhard, passed Kidd somewhere in the middle, and ended with Housman. "The ability to identify individuals at risk by DNA genotype promises significant clinical benefits." Perhaps *promises* was too strong a word.

The article quickly attracted a great deal of publicity. It made a charming story, with its visions of Amish countryside.

The Amish say "Siss in blut," Egeland told a *Time* magazine reporter: It's in the blood. "The next step for scientists will be to identify the particular gene or genes responsible," explained the article in *Time*. ". . . This will enable them to understand the biochemical basis for the disease, which could lead to better treatments."

The entire team—like many other researchers who have found genetic linkages—faced a long and uncertain wait. In the same issue of *Nature*, two other teams of researchers had reported that, in their own study populations, the linkage between bipolar disorder and chromosome 11 did not appear. One group, based in London, was studying an Icelandic population; another, headed by Gershon at the NIMH, had focused on ordinary North American families.

No geneticists impugned either the intentions or the procedures behind the Amish study. But two years later, no one had confirmed the linkage. Daniela Gerhard herself was no closer to the gene.

For all that it was one of the first DNA markers linked to a human disease, the chromosome 11 finding would join many other proposed but unconfirmed markers for the illness: the Xg^a blood type, inducement of rapid-eye-movement sleep, binding sites for neurochemicals on skin cells, and so on. It was an intriguing proposal. Was it more?

Many scientists decided that the gene might be a mutation peculiar to the Amish, their own unique way of inheriting this mental illness. Elliott Gershon gained approval for a large study to seek out other families that would be large enough to find the linkage again, according to computer estimates, even if it occurred in only 10 percent of cases. He felt there were two possibilities: either the Amish finding was wrong, or it hadn't been looked for enough.

"This project will look for it enough," he said. "If we don't find it, then it's not generally present."

In late 1988, in another issue of *Nature*, a team of researchers based in London reported finding a linkage to chromosome 5 in a set of large families permeated with schizophrenia and related mental conditions. The LOD scores ranged between 3 and 6, depending on the conditions of analysis.

Like the Amish paper before it, this one appeared with a companion report that failed to find the same linkage in a different kindred. The companion report came from Kidd's lab. Many months later, a separate study of Scottish kindreds would also fail to confirm the linkage.

The schizophrenia studies brought to the surface a deep disquiet within the young field of the molecular genetics of psychiatry. What was the explanation for the repeated failure to confirm genetic markers in psychiatry? Does it reflect some underlying diversity in mental illness? How much work would it be to sort that out? How soon would psychiatric genetics lead to real revelations rather than just more confusion?

There are still many among the Lancaster Amish who do not recognize the name Egeland. Say the word *depression* to an Amish farmer in Lancaster County, and he will most likely begin talking about an era before World War II.

But there are people in the county who remember Egeland well. One of them helped in her research, early on. Her blue eyes have the characteristic direct, unwavering gaze of most Amish women. She looks across her kitchen table, and remembers.

"Janice did everything she could to mimic our way of dressing," the woman says. "She dressed real plain, she wore a kerchief. She was real friendly. A lot of people just loved Janice; she made herself easy to love. A lot of people still do love her.

"I don't think about her much, haven't thought about her for a long time, in fact," she goes on, "until now. What's she doing now?" She waits a moment for a reply. "I know she has her hair all done up nice and she dresses up. I'll bet she's giving a lot of lectures."

Then there are others who know her still, and consider Egeland a very good friend. One of them is an unmarried woman of about thirty-five, to judge from her appearance, who works in a shop in a small town in Lancaster County and lives with her brother and his family on a farm. She is more than happy to take a lunch-hour break to talk about Egeland.

Let's call her Mary Stoltzfus, which joins a common Amish female name to the most common Amish surname. Before she met Egeland, Mary used to get "real low. Often I couldn't go to work. I just couldn't bring myself to do it." She shrugs. "It's a sickness," she adds quickly, and brightly.

The Stoltzfus family was happy to take part in the studies, after Egeland helped Mary to find psychiatric care, and lithium treatment, which has changed Mary's life.

"Now I can go to work, I can talk with my friends," she says. Janice drops by regularly, just for a chat, and sends Mary cards on holidays. "She lives for others," Mary says simply. "Oh, yeah. She's great."

Twenty-five years after the Amish first met her, Janice Egeland is now an ample, rose-cheeked woman in her fifties. She would remind you of your favorite aunt, or perhaps of a middle-aged nun. She has that same aura of purpose about her.

Egeland always presents a jovial aspect as she talks about her Amish study. She clearly loves to tell the story, and some elements are always present: drawing blood in the cornfield, explaining psychiatry by the light of a kerosene lamp, falling in love with the Amish. Audiences respond warmly and attentively. In the academic world, she has become a phenomenon.

Occasionally people refer to her as the Margaret Mead of the Amish. Egeland rejects the analogy. "I'm not like Margaret Mead, who left Samoa and came back twenty years later," she says. "In a way, I never left the cornfield."

In fact, as she pointed out when accepting an award from the American Public Health Association, now she can't leave it. The need to keep the information behind the cell lines up to date requires "the commitment to follow these families, shall I say to retirement—for me, shall I say to 2004?"

Meanwhile she has to hold at bay the reporters and other scientists who want to enter the community on the heels of her study. She is most concerned now to avoid jeopardizing her foothold inside a community that, she says, has begun to feel like the human equivalent of guinea pigs.

"I need to slow down, slow down, get back in the field," she said, during a long telephone call in late 1988. "Please, no more cameras, no more reporters. I need to be as unnoticed as possible, until, hopefully, there's another finding." For that reason, she went on, she could help me no further. Subsequent events changed her mind.

By mid-1989, there was another finding, but not exactly one she would have wished. At an international congress on psychiatric genetics, Egeland gave a companion report with a scientist from California, John Kelsoe, about a new set of findings based on the cells from the Amish. These were not nearly so promising.

Kelsoe and co-workers from the NIMH, initially fascinated by the idea of seeing whether a gene for a brain chemical on chromosome 11 was involved with the purported gene for bipolar disorder, had purchased some Amish cell lines from the repository in Camden. They had retyped the cell lines, tested them with some newer markers, and also incorporated some updated diagnostic information about people in the pedigree who had not developed the disease by the time the linkage first reported in *Nature* was established. Kelsoe's results matched what Gerhard was beginning to find, and the teams decided to collaborate.

After the updated analysis, the LOD score for the chromosome 11 linkage plummeted, from 4.7 to 2. Where people had seen a gene for bipolar disorder on chromosome 11, it vanished—at least until some further changes in diagnosis that might drive it up again.

"It was the biggest bombshell at the meeting," said one geneticist who attended it. "There's no question it's not significant anymore, but I don't think people think it's a bad study. It dealt with a lot of issues people weren't aware of, especially the matter of blindness and biases in the study design. Certainly the ones that are obvious they did compensate for."

On the whole, Egeland told me, the development was "very sobering." It implied that the hereditary basis of bipolar disorder might be both multigenic and polygenic—involving different genes in different people, and different genes in the same people. (On the face of it, in retrospect, logic tells us it ought to be this way. Any single dominant gene that could cause something as debilitating as bipolar disorder couldn't be very common in the population. Its possessors, after all, tend to eliminate themselves.)

As usual, however, Egeland was looking for positive messages from the disheartening turn of events. One was the absolute necessity, in genetic linkage studies for all diseases, to follow the subjects indefinitely, in order to know who did and did not develop a condition. One or two cases, she knew well, could alter the validity of a linkage.

The other was the importance of sharing, especially now. "If we don't feel that we can really share what we're finding with our colleagues, and maybe pool our efforts," she said, "we're not going to have as good a chance to get there. The positive message is that we might start working together."

Ironically, Egeland's own determination to share her cell lines and keep track of all her study subjects had reversed the finding she was so proud of. But no good scientist would have it otherwise. Nature may fool us, but we cannot fool nature.

6

OF COOKING, ART, AND MAGIC

In psychiatry, where the issue of heredity was almost taboo for more than a generation, the prospect of finding a gene that would explain a mental disorder was energizing. The newer reality, that several genes may underlie any common disorder, has proved baffling and disheartening. In other fields of medicine, researchers have taken that for granted for years.

In many such cases—diabetes, for instance—the nature of a disease has been clear for decades, parts of the basic biochemical defect are worked out in detail, and rational treatments are available. Still, the new genetics has a great deal to offer. The picture is not yet complete; newfound genes will fill in the missing elements.

Alberto Severini, his dark suit rumpled from a long day's work and two airplane flights, reached the lab in Edmonton at about 10:00 P.M. It was dark and silent. He flipped on the lights, and set down his burden, a large Styrofoam box.

His eyes felt heavy and his muscles ached, but Severini had no choice but to get to work. There were fifty vials of blood in the box. In a sense, the experiment would begin now.

Some people say Severini has a green thumb in the lab, but he generally thinks of himself as clumsy—especially that night, heavy-handed as he felt. He had to dilute the samples one by one, and then draw red liquid from each sample into the thin glass column of a pipette, using a rubber bulb. Then gently, gingerly, he needed to release the thinned blood onto the surface of a measured volume of Ficoll, a viscous fluid as thick as Coke syrup, inside a plastic tube.

That done, the tubes of Ficoll would spin in a centrifuge, paired to balance their weight, for twenty minutes. When he lifted them out afterward, the tubes would be striped: a blob of red blood cells at the bottom, beneath a wide gray stripe, the Ficoll, which was separated from the yellow-red serum at the top by a slender, milky layer of lymphocytes, white blood cells.

For the next two years, the lymphocytes in that thinnest stripe would be the principal object of his attention. In three months, when he completed his current fellowship in the biochemistry department, Severini would become the person chiefly responsible for the study of those white cells.

Severini says he chose research because in this field you can keep changing what you do. Since 1983, he had been studying ion transport in white blood cells. In 1989, he would be studying herpes viruses. This April night in 1987 he was preparing to analyze the DNA from families with a very strong streak of insulin-dependent (or, as it is misnamed, "juvenile") diabetes.

Tube by tube, fifty times, Severini had to tease the milk-white layer out from beneath the yellow-red one, again using a pipette. (On a much smaller scale, it was like drawing fat off the top of a soup with a basting tube—and then saving the fat and discarding the soup. But milliliters were critical.) Three times, he washed each sample of white cells in a solution that cleaned away the Ficoll "soup," spun the cells again, extracted them again from the Ficoll. Finally, he put the cells to rest in the freezer, suspended in fetal calf serum and other nutrients, as well as a substance that would prevent ice crystals from forming. The following night, another box full of blood samples would arrive, and he would start all over again.

It was the interminable end of a long day. At 8:30 A.M., Severini and two other members of the research team boarded a Time Air commuter flight northbound from Edmonton, Alberta. The small propeller plane droned for more than an hour over a thick carpet of treetops, and then came down onto a landing strip shaved out of the carpet: High Level airport, serving a row of truck stops and stores that cater to lumbermen, oil-field workers, and Indians from the nearby reserve.

Everyone on the team at this moment was a physician, working (like many doctors at university medical centers) in clinical research. The team would be studying people from La Crete, the same tiny farming town Roger Kurlan had followed David Janzen home to, more than a year earlier—but the Edmonton team had come for a completely different reason. Severini was not the leader of this research group; that distinction unmistakably belonged to Martine Jaworski, a pediatrician and diabetes researcher at the University of Alberta, who had gone ahead and would meet them at their destination.

Jaworski, ebullient as ever, was there when they landed, waiting with a rented car to drive them to La Crete. She had flown up a day ahead, to resume the long process by which a medical researcher adopts an entire community. She gave a Sunday-night seminar in La Crete, to explain to the families what the diabetes study involved and why they were being asked to help. She rose at 5:00 A.M. Monday and drove to High Level's little hospital to give an 8:30 seminar to doctors there. Then she drove to the airport.

Jaworski greeted the rest of her team with her lopsided smile and a chuckle. In deference to the Mennonites in La Crete, she'd asked her co-workers to dress very plainly, and she found the result comical.

Greta Mansour, the graduate student from Egypt who always dressed brightly to complement her olive skin and her personality, had to borrow a dark dress for the occasion. It made her look like a nun. Alberto showed up in his best outfit, a dark, well-tailored Italian designer suit. With his boyish, melancholy face, he looked like a lad on his way to his first Communion. Jonathan Slater, the pediatrics resident, stood between them wearing his usual businesslike air beneath his Arctic-weight parka.

Jaworski asked them to pose for a snapshot in the parking lot: Our Gang on a Field Trip. Then they drove off toward La Crete.

At first the highway was straight as a furrow, bordered by wide fields of grain that were rimmed by stands of aspen and willow. Then they turned to the south, crossed the broad and serene Peace River, and the smaller road began to turn and turn again, marking the edges of farms. So far to the north, it was surprising to see fertile farmland. Looking out the window, Greta Mansour wondered aloud: ''How did they find this place?''

Perhaps some of the first settlers had seen a 1920 surveyor's report to the Canadian government, stating that the writer had ''no doubt that this area would produce hay in profusion and that if . . . cleared and drained it would provide good land for coarse cereal.'' So it does: although the growing season is shorter than farther south, it is mild and sunny enough to grow wheat, oats, barley, and rapeseed.

The first group of settlers came up the Peace River by scow in 1934, some years before the area was officially opened for homesteading. It was one of many migrations for the Old Colony Mennonites, seeking yet again to be left alone.

When they came, northern Alberta was about as far as they needed to go. Even today, it's remote: On the few maps that show La Crete, it usually falls off the top of the Alberta section and onto the inset with the Northwest Territories, the region where Canada breaks up between rivers, bays, and straits, dissolving toward the North Pole.

La Crete was incorporated in 1974, and given the name used by three early French settlers who thought the nearby hills looked like a cock's comb. A paved road leads there now. Still, it's so far from anywhere that arriving makes you wonder why you bothered to go to so much trouble.

To all appearances, it's just another small town: a gas station, a café, a motel, a hardware store and a tractor dealer, two grocery stores, and, just before the highway shoots out of town, the small bungalow that serves as the public health unit. A sign in front says, in large letters, NURSE. Jaworski pulled up in front of it.

When she and Slater first visited the health unit a year earlier, in the spring of 1986, they were (as far as anyone knew) the first doctors from the nearest major medical center to set foot in the town. Jaworski was startled to learn from the nurses that a team of American medical researchers had preceded her there.

Roger Kurlan used the same public health unit, when he drew blood from the Janzen relatives to begin tracing the genes behind Tourette's syndrome. Jaworski's team and Kurlan's would work completely independently of each other, never crossing paths. When they compared pedigrees several years later, they would conclude that they had discovered two completely different sets of families, with two distinct hereditary ailments—Tourette's syndrome and insulin-dependent diabetes—in the same small Mennonite community.

But Jaworski did not stop with diabetes; she traced the roots of the community backward in time, and its branches outward geographically. One group among the Mennonites proved to be a most remarkable genetic isolate—at once withdrawn from society and dispersed across the Americas, from the northern reaches of Alberta to places as far south as Bolivia.

Where observation is concerned, Louis Pasteur said, chance favors the prepared mind. Martine Jaworski's mind was not merely prepared to observe a genetic isolate among the Mennonites in Alberta. It was primed, eager, lying in wait—perhaps, in a way, since her childhood.

A doctor's daughter, she carried her father's black bag to house calls when she was a child, and later she proofread his medical research papers on the subject of kidney disease. At least since medical school, she had been puzzling over the question of why some people develop complications after an infection, and others do not.

"I think I was at some elementary level toying with the idea that there was an explanation, some underlying pattern that would explain these variations," she said recently. "It's a viewpoint of the world as being orderly, predictable, things not happening at random." Well, things do happen at random, she corrected herself quickly, but in general there are explanations for nature. Science attracted her, she said, because "It's a search for underlying patterns, meaning, order. It ties things into a coherent whole."

Jaworski got her medical degree at Johns Hopkins, where she remembers Victor McKusick talking about the Amish studies and once or twice bringing an Amish patient for the medical students to see. His photographic portrait of an Amish mother stays in her mind. She took her residency at McGill University in Montreal, where medical geneticist Charles Scriver was studying hereditary metabolic disorders and other diseases among a different genetic isolate, the French Canadians.

Later, during a fellowship at McGill with endocrinologist Eleanor Colle, Jaworski began to study the immunological markers for insulin-dependent diabetes mellitus (often abbreviated as IDDM). This disease, which requires regular injections of insulin to maintain life, affects about one in three hundred North American children by the time they turn eighteen. Thirty or 40 percent of these will later develop severe kidney disease.

At the time Jaworski first focused on juvenile diabetes, in the heyday of research into organ transplants, immunologists were busily finding correlations between tissue types and various diseases. The correlations were neither perfect nor exclusive, but they were often tantalizing.

Jaworski was struck by the fact that children with two particular tissue-typing categories, HLA-DR3 or HLA-DR4, were about twenty times more likely than average to develop diabetes. "There was meaning and order there," she says, "but we didn't understand it at that time."

Jaworski's first academic appointment was at the University of Alberta, where an honorary foundation grant allowed her to spend three-fourths of her time in medical research. She began studying a breed of laboratory rats that develop insulin-dependent diabetes. She was methodically testing anti-immunity drugs on the rats, based on the widespread conviction that the disease is the result of an assault from within—the immune system, evolved to defend against infections, misguidedly storming one part of its own territory, the pancreas.

One day in late 1985, in her role as associate professor of pediatrics, she was asked to see a nine-year-old Mennonite boy with diabetes. She could not tell he was Mennonite by looking at him; that part of his identity came out later, when she spoke with his father.

Although Mennonites often turn up for treatment at the university hospital in Edmonton, a doctor would see no obvious reason to identify them as unusual. Except if you see them in groups in La Crete and notice how identically they are dressed, Old Colony Mennonites just look like farm folk. It's not especially surprising, perhaps, that for many years they mingled almost unnoticed within the Canadian health-care system.

A few days before Christmas 1985, Jaworski was asked to consult on the case of the diabetic Mennonite boy. Jacob is not his name, but it will do. Jakey had been airlifted to the University of Alberta medical center from a city far to the north, after an acute three-day episode of malaise, vomiting, and abdominal pain. As a pediatrician with a special interest in juvenile diabetes, Jaworski was often called in on such cases.

She met with the boy's father, a cattle and grain farmer, and ruled out exposure to a certain pesticide known to cause diabetes. What family information he gave her was vague, beyond the fact that none of his other ten children had ever had anything like this. He thought there was lots of other diabetes in the family, but he could not be specific.

A few days later, she had the opportunity to meet Jakey's mother. They sat across from each other in a small consulting room off the pediatrics department. Jaworski, a tall and slender woman with a commanding presence, must have been dressed as usual in a well-tailored skirt or suit, a classic blouse, and something bright for an accent, probably a scarf, certainly a white coat.

The mother was a forty-five-year-old woman with a quiet demeanor, a strong face, and the form of someone who bore eleven children and helped with the farm work afterward. She was wearing the costume dictated by her own circumstances: a flowered housedress, sneakers, a kerchief tied at the back of her head and secured with metal hair clips. The outfit is far less distinctive than what the Amish wear, but just as carefully prescribed.

The two women could watch Jakey through a window, as he played in the waiting room under the care of an older sister. "It looks like Jakey is much better," Jaworski remarked.

His mother must have had a vision of her son's being lifted on a pallet into the air ambulance. She was not allowed to accompany him on that flight. She had to return to the farm and wait for news from her husband, who had flown to Edmonton from High Level. When Jakey's father returned to La Crete a few days later, she took the bus down to Edmonton to stay with her son.

"The outlook is quite good once the diabetes is diagnosed," Jaworski said, "and as soon as Jakey can be treated with insulin. You do know that he will have to be on insulin for the rest of his life, don't you?"

"Yes, I know," she replied. "Jakey's cousin, Adam. He has it too."

"I'd like to talk to you some more about that in a moment," Jaworski went on. She always speaks in a didactic tone, which must be very reassuring to patients. Jakey's mother found her talk "plain and . . . just sensible," which from her is a high compliment.

"You know, juvenile diabetes is a familial disease," Jaworski continued. "It's one of the more common chronic conditions in adolescents. Maybe one of every thousand or two thousand children develop diabetes. . . . You know that Jakey's sweets will have to be severely limited now, don't you?"

"Yes, I know."

"I can imagine how difficult it would be to put one of my boys on a restricted diet," she remarked. "Will that be a problem for Jakey?"

The woman brightened. "You have children?"

"Yes, just the two boys. They're three and five. I see you have ten other children. Will it be a great problem for you to restrict Jakey's sweets?" she persisted.

"Well, some of the children are grown now. For the others, well—we'll do what has to be done." She fell silent, watching her son through the window.

"When I saw Jakey's father," Jaworski continued, "he brought up some very interesting points I'd like to ask you more about. He said there were other people in your family who have diabetes, but he wasn't sure who they were or exactly what they had."

She began to ask about Jakey's ten siblings, and the mother's pregnancies, and then the health problems of Jakey's aunts and uncles. The family was not Russian, Jakey's mother told her, correcting one of the few bits of information Jaworski had from her husband. They were German, she insisted, but they had come from Russia. They were Mennonites. The name meant almost nothing to Jaworski; she noted it on her sheet of paper.

The number of diabetics Jakey's mother knew about was surprising, to say the least: She named three of Jakey's close cousins as having juvenile diabetes, as well as another whose name she couldn't remember. A grandmother on his father's side, as well as three relatives on her own side, had diabetes, she said. During her own pregnancy, she had developed diabetes temporarily. In fifteen years of medical experience, and despite a habit of taking meticulous family histories, Jaworski had never heard of such a concentration of the disease in a single family.

As her notes stretched down the sheet of paper, Jaworski began to realize that she ought to widen her frame of reference in thinking about this unusual family. Again and again, she noticed, Jakey's mother alluded to relatives who had diseases that, like insulin-dependent diabetes, appear to be autoimmune. Rheumatoid arthritis, she noted on her record, and thyroid disease.

Maybe I should even look beyond the autoimmune diseases, she thought, beyond the family. She asked about inherited conditions and other diseases in the community, beyond Jakey's close relatives. Who were these people burrowed away up north in—now what was the name of that little town?

"I had more information than I knew what to do with," she recalls. "I didn't know how to organize it." After the interview, she thanked the mother from La Crete, and spent an hour writing a consultation report to add to Jakey's chart. But she resolved to write another "consult," in much more detail, as soon as possible.

There are those in genetics who like to portray the field as some sort of footrace or as a breathtaking adventure. To hear Jaworski talk (and she clearly loves to talk about it), genetics sounds more solitary and aesthetic, like cryptanalysis, or a game of chess, or something akin to art or music criticism.

Jaworski earned a piano performance degree during high school, and toyed with the idea of going into music. She often speaks about diseases running through a kindred the way a theme runs through a fugue by Bach. She also likes to compare the human population to a great tapestry or a mosaic, and likens her own work to an effort to discern the path of a thread in a tapestry, or a meaningful pattern embedded in a jumble of colors.

Christmas 1985 was a quiet time; Jaworski had many moments to muse about Jakey and his people. The more she thought, the more compelling they became.

"I was convinced there was more to it," she says, "and I was convinced that it was a genetic isolate. I didn't know what the degree of isolation was at that point; I didn't know where they were all coming from. All I knew at that point was that they had been in Russia, that they sometimes called themselves Russian Mennonites."

Jonathan Slater, a new pediatrics resident, was due to join the lab on New Year's Day. She decided not to ask him to work on the diabetic rat studies, but to ask if he would set off in search of Jakey's extended family. Slater began to telephone around, trying to confirm the names and diagnoses of people who appeared to have autoimmune diseases. Meanwhile, they set out to learn what they could about the Canadian Mennonites.

Jaworski went to the library and checked out McKusick's classic book about the Amish, to see how he had defined a genetic isolate. Slater was gathering the numerator in the fraction, the number of diabetics; she needed to define the size of the kindred or community, the denominator. Within the month, she flew to Ottawa to study the statistics of the Mennonite community in Alberta and in the rest of Canada.

During the visit, Jaworski learned that religion would not be used as a category in the 1986 Canadian census. If she was going to study the Mennonites, she would have to figure out the boundaries of the community for herself, somehow.

She began raising the matter at departmental meetings. "Nobody else from the faculty had set foot in La Crete," she recalls, "so I couldn't talk to anybody and say, how big is it, what's going on there?" In March she and Slater went to La Crete, to meet the local doctors and public health nurses, to seek out some genealogy.

Shortly before her first trip to La Crete, Jaworski telephoned the public health unit to introduce herself. She knew how close public health nurses are to a community in the Canadian medical system. They are responsible for most prenatal care, for new-baby checkups and school immunizations. They register everyone who turns sixty-five, see about routine care for the chronically ill, and even process death certificates.

Over the phone, she asked the nurses a question: Who in the community knew most about genealogy? The nurses sat down over coffee and thought about it afterward. The name they came up with, remarkably, was Jakey's own grandfather.

Jaworski and Slater visited him at the end of their first brief trip to La Crete, accompanied by one of the nurses. "He was extremely cooperative, not at all judgmental," recalls Jaworski. "He had a marvelous sense of humor. He had blue eyes that twinkled, and a sort of cherubic appearance. But he understood the import of all this, and he understood why we were asking what we were asking. If he didn't know the answer, which didn't happen often, he would tell us.

"In fact, when I saw him, I thought, this man has the natural mind of an empirical scientist. He hadn't gone to school, but he had this logical, rational mentality. And I could see why he had this role in the community."

In one long session around the kitchen table in his small cottage by the Peace River, Jakey's grandfather provided Jaworski and Slater with everything they needed to flesh out the pedigree. The following year, after his death, the family located among his possessions an ancient family tree, penned in spidery German script, that carried the relationships back as far as 1600.

In the end, the boundaries of Jaworski's isolate—her denominator—would define themselves historically, in surnames. They are the families descended from the original members of the Old Colony Mennonites, the most conservative of all the Mennonite groups in North America. Throughout the last 450 years they have distilled themselves, over and over again, until they show a distinctive pattern of hereditary disorders.

The Amish and the Mennonites come from western European stock. They split in 1693. Over the next century the Amish made their way to America, and isolated themselves within American society by their unusual life-style. The Mennonites had the same fundamental beliefs: adult baptism and pacifism. But they responded to threats from society in a different way. Again and again, they argued with their feet: they migrated.

During the 1700s, the Dutch-German Mennonites made their way to the Ukraine, having struck a mutually satisfactory bargain with czarina Catherine the Great. They could have religious freedom and a life of pacifism, if they farmed the land, refrained from trying to convert others to their faith, and if they practiced endogamy (marrying only among themselves).

This suited them perfectly. About four hundred Mennonite families founded a settlement, Chortitza, on the Dnieper river. There were later settlements nearby. The "Russian" Mennonites thrived for about a century, until a new regime reneged on the old agreement about military conscription, and insisted on taking over the Mennonite schools.

The Mennonites from Chortitza responded in their customary fashion. In 1874, they moved to Manitoba, where the Canadian government (as Catherine the Great had a century earlier) had an interest in settling vast lands and no disposition to make great demands on the settlers.

In his history of the Old Colony Mennonites, Calvin Redekop writes, "It is alleged that the more conservative and orthodox left Russia first. . . . If this is true, we perhaps can identify the motives of the members of the Chortitza settlements (the 'old colony,' since it was the earliest colony), who were among the first to migrate to Canada, as motives of preservation of the faith."

Even in Canada, the Old Colony members tended to be the most rigid of Mennonites when they saw threats to their faith, especially in the government's attempts to influence their children's education. Around the turn of the century, some Mennonites withdrew to the west, to Saskatchewan; twenty years later, another group resettled in Mexico. As recently as the 1960s, several very conservative Mennonite families from Canada migrated to South America, again to escape outside influences. The life there has been especially difficult, and many have migrated back, including Jakey's grandparents.

The settlement of La Crete in the 1930s was one of the last great migrations of Old Colony Mennonites within Canada, north and west far enough (as they saw it) to be out of reach of the school authorities and the dangers of the outside world in general. Ironically, however, in that remote corner of Canada within fifty years New York and Hollywood had invaded. The snack bar in the center of La Crete now rents videos and sells romance novels.

Because their farms are larger, or perhaps because they relied on remoteness to keep their communities together, the Old Colony Mennonites have not resisted many of the advances of recent technology (although they do look askance at computers, at least for now). They use tractors and cars, telephones and electricity. The contrast between two worlds is far more subtle in La Crete than in Intercourse, but it is there nonetheless.

There are four Mennonite churches in La Crete, one so liberal it holds services in English rather than German. To marry in an "Altkolonie" church, both parties must be members. Like the Amish, the Old Colony Mennonites also tend to have large families.

"The practice of endogamy, dating back to the mid-sixteenth century in Europe, has been continued by the more conservative branches of the church," Jaworski and her coauthors wrote in an article in the *American Journal of Medical Genetics*. "It has perhaps been accentuated in Canada by the geographic dispersion in small rural settlements, and by splits among church branches."

As a result, the Old Colony Mennonites have probably begun to exhibit two classic phenomena of evolution: *founder effect* and *genetic drift*. Because they trace back to fewer than four hundred couples, by definition the genes being shuffled between today's Old Colony members are limited to those passed down by the founders of the sect. Some alleles (variants of genes), including many that cause diseases in the general population, are doubtless absent in the Old Colony Mennonites. A few that do cause certain diseases are probably unusually common in that community.

At the same time, random fluctuations in the sorting of genes during reproduction, as well as the disappearance of some genes due to premature death or infertility and the inevitable effects of endogamy, have doubtless led to genetic drift, the proliferation of some genes and the elimination of others. Just as isolation on an island can lead to unique species—the koala bear, the kangaroo, and the duck-billed platypus in Australia, for instance—the condition of autoimmunity might be accentuated in the Old Colony Mennonites.

A few days after she met Jakey, Jaworski wrote a more detailed consultation report for her colleagues at the university and for Jakey's doctor in Fort Vermilion, the nearest large town to La Crete. She thanked them for referring an interesting case, and added her opinion that the family, and indeed the community, were worthy of careful study.

By that time she had a clear enough idea of what she meant by that to specify her plans. She had frozen Jakey's serum to look for antibodies against his pancreas cells, signs that his immune system was attacking his own machinery for insulin manufacture. She had deferred plans to test Jakey and his entire extended family for their HLA types, for the tissue-typing categories that tended to coincide with juvenile diabetes. Jakey, she predicted, probably had the HLA-DR3 or HLA-DR4 tissue types, or both.

"Again and again," reads a quote from an article that Jaworski sent, unsolicited, to help prepare this chapter, "the record reveals that the discovery is not a fluke but the inevitable, if unforeseen, consequence of a rational and carefully planned line of inquiry initiated by a scientist." She wrote the word *true* in the margin in red, and underlined it.

As Jonathan Slater was busily tracking down Jakey's kindred, Jaworski had another immediate concern: She was to be cochairman of an international conference in Edmonton during the summer of 1986. The subject was the immunology of insulin-dependent diabetes.

It was abundantly clear that the disease was an autoimmune disorder. Laboratory tests almost invariably showed that people in Jakey's situation had begun to produce antibodies, the molecules of immune recognition, which targeted their own beta cells, the cells in the pancreas that make insulin. These antibodies were hardly predictive of diabetes: They were a sign that it was too late. The body had already begun to destroy its own capacity to manufacture insulin, the hormone that helps it make use of sugar and other carbohydrates.

What was urgently needed was a way to predict who was

on the road to juvenile diabetes long before the kind of crisis that had brought Jakey to Edmonton. The DR3/DR4 HLA classification Jaworski alluded to was unsatisfactory for this purpose. What it stood for was merely a general class of molecules on the surface of cells, molecules used by the immune system as a sort of passport, to distinguish what belongs inside the body and what doesn't.

Just as few international travelers are terrorists, only a small subset of people whose cells carry the DR3/DR4 "passport" have immune systems that for some reason decide to attack their beta cells. The imperative was to find a way to identify that subset, and the obvious strategy was molecular genetics. Luckily for diabetologists, there were two analogies in the animal world: the BioBreeding diabetic rat and the nonobese diabetic (NOD) mouse; both strains developed syndromes apparently similar to juvenile diabetes.

By the time of the conference, two teams of scientists (headed by Hugh McDevitt at Stanford University and Gerald Nepom at the Virginia Mason Research Institute, an immunogenetics research center in Seattle) were working intensely and in parallel to try to identify this subset in humans. By extrapolating from genetic studies in the mouse, Nepom had identified an important category of DNA fragments within the genes on the short arm of chromosome 6 that code for the HLA-DR4 type. It was a subclass of the DR4 tissue type, known as DQ.

"A striking correlation of the DQ3.2β [pattern] with IDDM is apparent," Nepom reported at the diabetes conference in Edmonton. ". . . Over 95 percent of DR4-positive IDDM patients analyzed in this way express the DQ3.2β allele." In diabetic mice, McDevitt's group had found an even finer distinction: a change in a single amino acid, a single unit in a chain of protein building blocks. The fifty-seventh amino acid in a protein molecule on the surface of certain immune cells—the molecule that determined their DQ tissue type—was different in diabetic mice. This change in an amino acid was the result of a distinct molecular variation in the corresponding mouse gene. A year later, McDevitt's team would report that many humans with IDDM have a mutation at exactly the same spot.

At the Edmonton meeting, Jaworski (who had already been in contact with Nepom) discussed with him her plans to trace diabetes genes in the Mennonite kindred. By the end of the conference, Nepom had agreed to collaborate: He would send her the probes, the DNA fragments that could point out the DQ3.2β pattern, and she would try to confirm the linkage he had found.

On a warm Monday the following April, having recruited her research team and enough funds to get the project started in earnest, Jaworski descended on the La Crete nursing station with her entourage. She had planned the trip for just before plowing season, when the farm families would still be relatively free during the days.

All winter—in fact, since the previous year—Jaworski had been calling the unit at quite regular intervals with varying requests for this or that, and the nurses began to moonlight for her. They scoured their memories for anyone among their patients who needed to take insulin; knowing them personally, this was faster than looking through the files. They tracked down these people and their relatives, phoned them, briefly explained Jaworski's purpose, and set appointments for them to come in for interviews and blood sampling during three days in late April, just before planting season.

In the meantime, also for the energetic lady doctor from Edmonton, they had been fingering through their own records and calling local doctors to document the extent of diabetes and a host of other ailments. The week before the team arrived, they rearranged the furniture and equipment in the health unit, setting off private areas where they and the four doctors could meet with the visitors.

The first of the families arrived before 1:00 P.M., and the work began on a tight schedule, fifteen minutes per interview. Slater and Severini drew blood, the nurses completed questionnaires, Greta Mansour spun the samples in the centrifuge and put them in the freezer.

On the first day, the last visitors left after 5:00 P.M. Mansour and Severini took the vials from the freezer and plopped them one by one into the wells in the Styrofoam box. Someone gave Severini a ride back to High Level, where he boarded the little prop plane again. It landed in Edmonton about 9:00 P.M., and he went straight to the lab, where he worked until 4:00 A.M. preserving the samples.

Jaworski began early each morning, with breakfast-hour seminars, and worked straight through each day interviewing family members and taking genetic histories. Distances are great in northern Alberta; with the seminars and trips to the airport, she got in a lot of driving.

She closed the clinic down every evening, interviewing families. One evening she spent a long time consulting with the cleaning lady.

For many months, Jonathan Slater had been principally occupied with tracking down Jakey's relatives, searching for others who had juvenile diabetes and talking with their doctors, tracing Jakey's family tree, and learning his ethnic history in intense detail. All of the people giving blood in the clinic for three days were relatives who had been traced in a sinewy line extending from Jakey. Suddenly one afternoon Jakey himself materialized in front of Slater.

Slater shook Jakey's hand and asked how he was doing. The answer was obvious from his robust appearance. Slater was conscious of the impression that something about the moment was too ordinary.

"Here's this kid around whom all this activity has been organized and revolved," he mused later. "I meet this kid—nice kid, nice parents. Like any other kid." Whatever the headlines may say, the progress of science is marked by very few eurekas, and by countless such moments that seem, somehow, too ordinary.

Jaworski and Slater were the last to leave La Crete. Three days after they began drawing blood they packed up their goods and disappeared. Jaworski was exhausted, but Slater felt fine. As a resident on call every two or three nights, he regarded the La Crete field trip as a small vacation.

The nurses waved good-bye, tired and a little let down. When they were gone, "it was like a big hole," said one.

That was hardly the last they heard of Martine Jaworski, however. She returned the results of the study to La Crete, with instructions that the nurses should send them to each family's doctor. She asked them to help her verify the updated pedigree. She continued to phone asking for information, and regularly sent them fliers and medical reports on a wide range of topics.

Jaworski had been tracing the Mennonites to other parts of

the continent and trying to survey the entire spectrum of their illnesses. Gradually, she would submit the information to journals sent to doctors across North America. Her intention was to help them identify the special health problems of Mennonite patients, soon enough to treat the illnesses promptly or even prevent them. She always offered to provide a list of Mennonite surnames upon request.

Usually she accounts for this in terms that echo Victor McKusick, as returning something to the community. But there was another rationale too.

The genetic study of diabetes "is fairly high-risk work," she once said. "We had no guarantee that it would pan out to anything. So at the same time we had the opportunity to define and study a genetic isolate, quite apart from diabetes, and we decided to do that as well, because that was very likely to yield results. . . . You wanted something practical to show for all this work."

The truly risky part was Severini's business: transforming the cell lines so that they would grow indefinitely, and then testing them against Nepom's genetic marker. He began full-time in July 1986, and two years later had only partly finished. Severini is by no means slow or incompetent; he's just in a chancy profession.

By the end of that week in La Crete, there were five hundred little tubes in the freezer, waiting for Severini to find time to make them immortal. In principle, he looked forward to the work. It would be a fascinating contrast to the biochemistry he had been doing.

At the time the cells arrived in Edmonton, he had another task waiting: the data analysis for his current project, about ion transport across cell membranes. The data were numbers of debatable validity, measurements made under various conditions he had set up to approximate what he *thought* to be close to the situation inside the body. They needed to be analyzed by taking into account, insofar as possible, what might have happened with regard to the thousand other factors in the cell's normal environment that he had been unable to measure at the same time.

The raw data from La Crete, however, were not numbers

but families—tangible, three-dimensional individuals conjoined by rows of lines on a pedigree. They were people who either needed insulin injections to survive or got along just fine without them.

The patterns of inheritance on the pedigree were most confusing, to be sure. But assuming all went well, the patterns of radioactive bands he would derive from the family members' cells would be as unmistakable as traffic lights: either a cell had certain DNA fragments, or it did not.

The first order of business was to cultivate these cells in the freezer. This was a mere procedural detail, a matter of materials processing, not even worth a mention in the final report. It would have taken a trained technician a few weeks to do this, but for a new pilot study Jaworski didn't have funds for a trained technician. Severini had never transformed a white cell before, and the process consumed three frustrating months.

Transformation is the process of altering cells to a precancerous state, so that they will survive and multiply in a laboratory dish.

On the face of it, transformation is like baking bread. He had to gather and measure the ingredients, combine them, warm them, and wait for the cells to "rise." In fact, transformed cells on a culture dish do puff up, not unlike minuscule white muffins. Of course there are recipes for this procedure, in scientific papers and even whole genetics "cookbooks." But there are little details any good baker knows about that never get into a recipe. Lab work, like bread making, seems to be part technique and part magic.

What's more, no baker has to work with poisonous yeast, but the viruses that transform human cells are hazardous. They are in the same class as viruses that have been implicated in herpes and even cancer; any virus that makes human blood cells grow indefinitely in a lab dish deserves a certain respect.

The viruses arrived in Severini's lab from a commercial supplier, in purified form. A baker tears open a yeast packet; Severini had to layer a solution containing the viruses over a culture dish of white blood cells.

White blood cells are the scavengers of the immune system. Researchers use them for genetic research because they are relatively easy to collect and, unlike many human cells, they grow well in tissue culture after being transformed—turned, if you will, halfway cancerous.

Of course Severini did not begin practicing on the hard-won cells in the freezer; he went to a nearby lab and asked for volunteers willing to spare some blood cells so he could try to transform them. He exposed these to the virus solution, and waited. They sat in the culture dish, and sat, as lifeless as baker's yeast that fails to foam.

It took several tries for Severini to decide that the problem was not his but the viruses'. So he borrowed another supply of virus from a lab in the biochemistry department, and many weeks later than he had hoped, Mansour brought him a culture dish containing little white clumps.

At last he took some cells from the freezer and exposed them to the virus. Almost as soon as they started to grow, he noticed a greenish-white film on the culture dish: mold. Fortunately he had taken only a sample from the freezer, not the entire batch of a certain cell line. Nonetheless, it was disheartening. Severini dumped the contaminated dishes into a trash bin and sank into a deep depression.

Today he dismisses the mold contamination as almost inevitable. For lack of space, the incubator was in a different room from the biohood, the enclosed lab counter where the scientists could manipulate cells without fear of contamination from the open air. "We had to carry plates back and forth in the hallway," he recalls. "It's the sort of thing that happens when you're setting up a lab."

Even so, that was small consolation for the fact that he had everything he needed for a fascinating experiment, and could not even get it started. For several days, Severini backed off. He was even more quiet than usual. He mixed up batch solutions of reagents, got down to some paperwork, ordered supplies.

"I got very depressed," he concedes. "But that's research. It happens quite often that things don't work. Then you solve the problem, and the depression goes away." The respite dispelled the dark magic; the next batch of cells out of the freezer made it safely to the incubator, and five months after the blood had been drawn, the cell lines went into production.

British physiologist Bruce Charlton wrote about magic in a half-serious essay in the magazine *New Scientist*. ''If you want good results always wear the favorite tie,'' he wrote, ''but not the training shoes; mark the test tubes with blue ink . . . only use the glassware with the red masking tape labels. . . . Each lab has its own distinctive foibles. . . .

''There is no sharp borderline between magic and science,'' he concluded. ''Both depend upon fixed ritual to produce a predictable result. The only difference is that in science we think we know how the ritual works; in magic it works, but we don't know why.''

Severini's ritual centered on the probes contributed by Nepom and later by J. W. Yoon at the University of Calgary. It's a good piece of jargon, the word *probe*, with its connotation of reaching into unseen places and feeling about. A DNA probe is a small piece of genetic message that seems to search through fragments of DNA extracted from a cell line until it finds a matching genetic sequence, and then locks into it, in jigsaw-puzzle fashion.

Severini's probes were DNA fragments that could identify the DQ-beta segments, stretches of the gene coding for the tissue types that distinguish healthy people from those whose immune systems attack their own pancreas cells. A restriction enzyme chops DNA differently depending on the tissue types it represents, and thus creates fragments whose sizes differ by tissue type.

The probes arrived in Edmonton incorporated into *plasmids,* ringlets of DNA. These minichromosomes were discovered in bacteria, quite separate from the rest of the chromosomes but reproducing just as faithfully. In nature, bacteria tend to package genes for antibiotic resistance inside plasmids. They will also take up unrelated genes from totally different species. In the last fifteen years or so, biotechnologists have been slipping a tremendous diversity of other genes into plasmids, for their own various purposes.

Using the plasmids is most risky. The experiment, involving radioactivity and chemicals, can endanger the researcher. And the researcher, a walking ecosystem of biomolecules and microorganisms, can contaminate the experiment. It takes about a week to test a probe on a single cell line, many months to look for a meaningful pattern in an entire pedigree, and a lifetime for a scientist to suffer the effects of his exposure, if he ever will.

Before he could use the plasmids, Severini had to increase his supply of them. This is not very tricky: all he had to do was to grow them up inside bacteria. Later, he had to break the probes away from the plasmids using enzymes, and then make them radioactive.

Doing so, he embraced a large Plexiglas shield that rose from the laboratory bench plastered with yellow warning stickers. He measured drops from a pipette into Eppendorf tubes, tiny plastic vessels roughly the size of a ballpoint-pen cap fitted with little lids.

The Eppendorfs contained a solution of DQ-beta fragments he would use for probes. Inside the pipette was a mixture of the four DNA nucleotide bases, one of the four made with a radioactive element. After measuring drops into each of the Eppendorfs, Severini set the tubes in a rack, and went on to something else. Nearby, a Geiger counter crackled idly.

Inside the Eppendorfs, the bacterial enzyme began assembling: adenine matches thymine, and if a C is here, then add a G, and so on down the chain. When the reaction was complete, it would have produced an Eppendorf tube full of molecular mirror images of the probe, each of which would bind to any complementary, or corresponding, segment of DNA from the Mennonite blood samples, according to the rules of the genetic code. On an X-ray film, it would flag any sequence that was genetically equivalent to the probe.

Severini moved away from the Plexiglas shield, stripped off the surgeon's gloves he had worn to protect his hands from radioactivity, and discarded them. Then he put on another pair of gloves, to prevent him from contaminating the next job. While he waited for the enzyme to assemble his labeled probes, he began to prepare a blot. This is the small nylon sheet that picks up fragments of DNA from a gel on which they have assorted themselves by size. *Blot* is a good term, too, because the nylon picks up DNA much the way a blotter picks up ink.

Severini's gels contained DNA fragments from different Mennonite cell lines, which had spread out along their lanes like horses on a racetrack; the smaller fragments, in general, move farther in the presence of an electric field. Eventually a blot will soak up some of the DNA in exactly the same pattern in which it has spread across the gel. Each spot represents a fragment of unique size. (See the illustration on page 229.)

Like the vast majority of labs, Jaworski's team made their own gels. Often, the gel-making job fell to Mansour. It's a lot like making Jell-O, except that the powder is agarose, not gelatin, she used an automated stirrer rather than a spoon, and after the solution cooled she poured it into a flat tray and not a bowl. The resulting gel is a thin slab as stiff as wax, not at all jiggly.

Severini or Mansour would drop the DNA fragments extracted from their blood samples into small wells at one end of the gel—one well per cell line—and flip a switch. Overnight, in an electric field, the fragments would crawl through the gel in a straight line, one lane for each person's DNA.

That pattern, representing the sizes of all the DNA fragments from the cell lines of several individuals, blots up intact from the gel to the filter. The sizes of the fragments can be calibrated by reference to a standard, a sample of fragments whose sizes are known, run on the same gel. After soaking the filter in solution with the radioactive probe, Severini would make an X-ray image of it. The image would contain an array of ladder-like strips, side by side.

If a radioactive probe flags a different pattern of spots in DNA from two different people, this indicates that it has detected a *restriction-fragment polymorphism*—a variation in the sizes of fragments created by restriction enzymes that cut DNA, due to a variation in the DNA sequence at some point in the human genome. On an X-ray film of the blot, if such a variation exists within a set of those cell lines, some rungs of some "ladders" (or in geneticists' jargon, some *bands*) will be in different positions.

People searching for new genetic markers hope to find such a variation using a new probe. Severini already knew the difference existed, and knew how to detect it using a known probe. He merely wanted to see if the polymorphism would correspond with the presence or absence of juvenile diabetes in his particular population of subjects, the Old Colony Mennonites.

Altogether, the method is called *DNA hybridization*, the same technique Ken Kidd learned in order to test the Amish blood samples, and that Susan Chamberlain used to find the Fried-reich's ataxia linkage. At that moment, certainly, geneticists all over the world were going through the same motions: spreading bits of DNA across gels using an electric field, blotting them onto filters, lighting up the DNA fragment of their particular interest with radioactivity.

For their own purpose, Severini and Mansour had to repeat these steps and many more, hundreds of times, testing scores of cell lines against the two DQ-beta probes (and later, others). All the while, as is always the case in this singular kind of cooking, they had no idea what the final result would be. They had the recipe, but they couldn't say exactly what they were making.

The work with the DQ-beta probes was complete by the spring of 1988. Two years after the bloods were drawn, Severini and Mansour had established that Jakey and all of his diabetic relatives had the same DQ3.2β genetic sequence that Nepom had found in other "juvenile" diabetics. So did some other people, who were presumably at risk of developing the disease later.

The result was satisfying, of course, but it was hardly the end of the experiment. Of itself, at best, it helped to confirm a genetic linkage to insulin-dependent diabetes. For a few families in a small farm town in northern Alberta, it gave advance warning that certain individuals were at risk.

In May 1988, Jaworski got a telephone call from La Crete. She didn't recognize the name, among all the others on the pedigree, until the man identified himself as Jakey's father.

"One of my other children has come down with diabetes," he said. "I know you've done some studies on the family. Can you tell me who it is?"

Jaworski put down the phone and went to the files. "There were four children in the family who shared Jakey's [HLA] type and also had [antipancreas] antibodies," she recalls. "I said, I can't tell you which of these four, but it's going to be among these four, if our tests are correct." They were. The four included Jakey's affected brother.

"He was testing us," Jaworski said a few months later. "He also wanted to know whether there was any way of preventing it. I said, At this point, early diagnosis is the best thing. . . .

Right now, there is no practical application, other than suggesting both to the family and the physician to watch out for the really early signs of diabetes.

"But there may in five years' time be an application," she went on. "I think it might take the form of immunosuppression, intervening before there's full destruction of the pancreas. The diabetics in this family may get kidney complications and blindness, so we're talking about a serious disease. If there's any way to arrest the process before the end-stage destruction of the beta cells, we would like to do that."

Nepom's studies of Jakey's blood showed that he does possess the same DQ3.2β variant that McDevitt found to be common in insulin-dependent diabetes; he also possesses several other variations at other sites in the protein sequence. Nepom's studies, progressing since the Edmonton meeting, support the idea that these molecular alterations work together. In the shape-matching language of biology, where molecules communicate by touching and fitting together, they somehow send an erroneous message within the immune system, which sends it chasing after the beta cells, in error.

That can't be all there is to the story. The statistics alone tell you that: if someone has the identical HLA pattern as his sibling with insulin-dependent diabetes, he has only a 20 percent or 30 percent chance of developing the disease. If the two siblings are identical twins, genetic carbon copies, the risk is still only about 30 percent.

Because identical twins are genetic copies of each other, the latter statistic implicates some environmental trigger, as yet unknown, in the onset of juvenile diabetes. The fact that an increased risk exists for nonidentical siblings implies that more than one gene must be involved in susceptibility to the disease. The ill sibling has them all, and the well one, presumably, only the HLA gene.

Animal studies of diabetes have suggested a number of other candidates, primary among them the gene that encodes another molecular signal in the immune system, a cell-surface molecule that occurs in a class of white blood cells known as T cells. Through structures on their outer membranes, T cells form complexes with HLA molecules on the surface of other white cells. Together, they bind with molecules on the surfaces of entities that should be destroyed, such as pathogens—as a catcher might grab a softball in two cupped hands.

Genetic studies of other families with insulin-dependent diabetes had yet to provide any clear evidence implicating a T-cell receptor gene in susceptibility to IDDM; Jaworski and Severini hoped that the special nature of the Mennonite kindred might give an unmistakable result. When Jakey's father called, Severini was testing cell lines from Mennonites with diabetes against a new probe for the T-cell receptor.

By midsummer, when Jaworski returned from a vacation, he had the result they had hoped for: with one exception, all the people with insulin-dependent diabetes tested positive for the T-cell probe. "That's great," said Jaworski. "Now let's see about the rest of the cell lines." The result would be interesting only if most unaffected people tested negative.

The T-cell receptor studies are intriguing, too, but Jaworski is after a hypothetical set of genes, or perhaps a single gene: one that confers a general susceptibility to autoimmune reactions. One genetic trait may generally predispose the immune system to respond to what should escape unnoticed, she suggests, and other traits predetermine whether it attacks the pancreas to create diabetes, or the lining of nerve cells to create multiple sclerosis, or the lining of the joints to produce rheumatoid arthritis, or any of a number of other tissues, or none at all.

The search for such an "autoimmunity gene" might take many years, because by definition no one has any probes for it. In fact, no one knows that it exists. Will Jaworski and her coworkers or someone else begin the arduous effort to chase it down? It will be fascinating to see.

"It was heretical to think that such a multifactorial disease [as diabetes] could come down to such a reductionist approach," Jaworski reflects. "But molecules don't have any imagination. They're not programmed to do more than one thing. Therefore we should be able to see what they're doing."

7

THE ACCIDENTS CALLED CANCER

In the cells of generation, sperm and egg, genetic errors can become part of an inheritance, and leave a legacy of disease. Genetic mishaps can befall other cells in the body, too. The legacy of those mishaps descends only to new generations of cells, not to a person's sons and daughters and grandchildren. The result, nonetheless, can be a catastrophe.

Ten minutes to nine of an autumn morning: Raymond White was lecturing. He paced back and forth at the front of the auditorium, in green safari trousers and tweed jacket, his eyes scanning the audience.

The subject was cancer. At the moment, he was speaking about familial polyposis, one of the primary subjects of research in his own laboratory. The condition, which is rare and clearly hereditary, causes multitudes of small benign growths to arise in the colon, a part of the large intestine. Left alone, they inevitably lead to carcinoma of the colon—cancer.

"Extreme action is required," he said dryly. "The extreme

action in this case is colectomy. They remove the colon. It works, but it's a little *draconian.''*

Members of his audience listened impassively as White paced; their eyes wandered, they yawned and passed notes. White, who is among the most eminent geneticists of his generation, doesn't need to go through this. In fact, a co-worker told me, it even frightens him a little. But he feels a sense of obligation about it.

It was a classroom of undergraduates at the University of Utah, where White works. If there's a world-class team of geneticists right up the hill, he figures, the undergraduates deserve a firsthand progress report, something beyond the textbook.

''Despite all the well-founded criticism of the war on cancer,'' he went on, with all the gravity of a congressional witness, ''war or no war, a detailed picture of molecular events involved in carcinogenesis is starting to emerge. . . . The classic hypothesis used to be: Cancer, is it genetic or environmental? It's more and more clear this is not a dichotomy.''

The recent revelations about cancer, from White's lab and many others, have been astounding, enlightening, and portentous. It's ever clearer that cancer is the result of a sequence of genetic events, a chain of successive accidents involving damage to DNA. Some of these accidents are hereditary; others accumulate during a single lifetime, the insults of environmental damage—of radiation or chemicals or viruses. The genetic revolution has made much of this plain; the same techniques now give us the tools to understand those events more fully.

Ray White developed some of these tools, and he is energized by the fact that they can now be used against cancer. This comes across clearly when he speaks to his peers. He can seem larger than life, charismatic. He moves and talks in fifth gear. For some reason he failed to convey this to the undergraduates after several hours of preparation, but it was obvious enough when he responded, on the spur of the moment, to my question about his major research interests. We were walking across a lawn, during a break in a genetics conference.

''The object, once you've developed a set of tools,'' he began, ''is to find a way to make them useful that is the most . . .

fun. And it seemed to us that some of the best uses for these tools would be the study of human cancers.''

The tools developed in White's lab are thousands of probes that can reveal otherwise invisible flaws in a chromosome. For him, *toys* might be as good a word as *tools*. Cancer is many things to many people, but it's evident that to Ray White it represents the best kind of game in the whole world. It's a mystery whose characters—genes—he knows fairly well, and new clues are coming to light constantly.

Another scientist in White's lab, Mark Leppert, is assigned to maintain contact with the family members who donate blood for the lab's colon cancer studies. White himself avoids meetings of charity groups, Leppert told me, and other occasions where he is likely to be accosted by patients or their families, appealed to for his attention or urged to change course. Leppert defended this policy.

''You can't just say, these people need help, why don't we go out and start working on them?'' he insisted. ''You have to have a program. You have to have the experiments. You have to believe that you will get somewhere.''

I do not mean to imply that Raymond White is not deeply committed. He *is* deeply committed—to solving the mystery. Why different people choose to investigate the genetic mysteries of cancer is merely a curiosity. That they solve them is essential.

The person in White's lab who actually scrutinizes the DNA in search of a gene responsible for familial polyposis, as a clue to colon cancer in general, is Yusuke Nakamura. He comes from Japan, where until recently colon cancer was quite rare.

Since World War II, the major known dietary cause of colon cancer, fat, has more than doubled in the Japanese diet, as the Japanese have begun to adopt more Western eating habits. It's still not entirely clear how fat contributes to colon cancer; nonetheless, the shift in the Japanese national diet has been accompanied by a striking rise in colon cancer. In our first conversation, Nakamura mentioned how popular fried chicken was becoming back home.

Cancer is especially difficult to cure in Japan, where people tend to believe that it cannot be fought. ''Most people believe that cancer equals death,'' stated the head of the hospital at

Japan's National Cancer Center, shortly after Emperor Hirohito died of pancreatic cancer. "That is, all cancers equal death. So patients are often not told."

Yusuke Nakamura practiced as a surgeon for four years before he decided to switch to molecular genetics. "Basically, I was an abdominal surgeon," he told me, "so I did operations with gastric cancer patients, colon cancer patients. I had one colon cancer patient, he was a most impressive patient for me, because when he came to our hospital he was almost at the end stage of his cancer. Basically, a big mass of colon cancer obstructed his colon.

"He was very young, in his thirties. He had a wife and two small children, and probably he realized what it is he had." He could feel the mass in his own abdomen, Nakamura said. "He couldn't eat anything and he continued vomiting. So we knew it was end stage.

"At the beginning we tried anastomosis, a kind of [surgical] shortcut. When we opened his abdomen, he had a big metastasis in his liver. Anyway we did the anastomosis, and that operation was doing very well, and he would start eating again."

Unable to tell the man he had cancer, for cultural reasons, Nakamura told him instead he had been suffering from an infection of the bowel. The patient survived for another two months.

"The day when he died," Nakamura went on, "basically we knew he's dying. So I came back to my house. I wondered, I should go at the last moment, or not? I wondered. I may cry because I know his wife and small children were there. . . . I thought for several hours, I should go, or not? And at midnight one nurse called me, and his blood pressure goes down. I still thought, I should go, or not? I'm not sure I—I can watch. That is not the first time a patient is dying, but I have a very hard time to tell, your husband died."

Did you decide you should go? I asked.

"Yeah, so—" he cleared his throat. "So, maybe one hour later, [when] I arrive at the hospital, he died."

Did you cry?

"Not at the bed. I cried in my mind." Nakamura stayed two or three minutes longer—enough to inform the wife—and then he left, to avoid seeing the children crying. "I still remember seeing, he died," Nakamura told me. "Maybe that is my motivation for working hard," he added.

About a year later, Yusuke Nakamura decided to change careers. "At that point," he said, "until very recently, I am not sure I should continue research or I should come back as a surgeon. Because I didn't have much inclination, as a scientist, until recently."

By this point, however, Nakamura was quite definite that he would not go back to Japan to become a surgeon again. "Only because now I am very close to the polyposis gene," he explained. "Also, I am interested in an anticancer gene for colon cancer." At the end, a surgeon can do little to stop colon cancer, however deep his compassion. A molecular geneticist might just learn how to arrest it at the start.

Ray White's involvement with colon cancer traces back to the late Eldon Gardner, who was a biologist on the faculty of the University of Utah. One day in 1947 Gardner remarked to an undergraduate class that he had joined a research project to study inherited cancer. After class, a student came up to Gardner. "I know a good family for you to visit," he said. "They all die of cancer."

The next weekend, Gardner drove to a small town south of Salt Lake City to visit the young man's neighbors, a Mormon family, and he found it to be true. All the deaths in the family in the last twenty years had been due to cancer. Eight members of one family group had died of cancers of the lower digestive tract.

Gardner decided to trace all the descendants of the earliest family member certain to have had cancer, a woman who had died in 1909. He set up a clinic at the University of Utah Medical Center, and began to bring members of what he called kindred 109 to Salt Lake City for examination, transporting them hundreds of miles in his own car, weekend after weekend. This was no matter of mere bloodletting: among other procedures, they all had to be examined by colonoscope, a rigid instrument that was inserted up into the digestive tract from the rectum.

Of the first 51 family members Gardner brought in, doctors

found that 6 had many benign growths called *polyps* of the colon. Two of the 6 also had cancer in the same region.

By 1951, Gardner had examined all but two living members of the family over the age of five. He published a medical report of his findings, describing a hereditary pattern of polyposis of the colon, which seemed inexorably to progress to cancer.

"The pattern of inheritance of polyposis shown in this group is characteristic of a simple dominant gene," Gardner wrote. In every case where a child is affected, he said, so is one of the parents, and no child of two unaffected parents ever developed the syndrome.

In addition, many members of the family developed conspicuous lumps and benign tumors of the face and head. In the coming years, the complex of symptoms became known as Gardner's syndrome.

Numerous physicians had also noted intestinal polyps associated with other rare conditions, and it became difficult to distinguish all these syndromes from each other. By the 1980s, it was common to lump them all together under the name *familial polyposis*. The precancerous condition is essentially curable, if the colon can be removed before the benign polyps called adenomas turn to carcinomas, invade below the surface of the intestinal wall, and spread elsewhere in the body.

Hereditary polyposis syndromes of the kind Gardner studied account for only about 1 percent of the colon cancers in North America. The nonfamilial or sporadic forms of colon cancer are exceedingly common, however.

In every case, apparently, they also arise from an intestinal polyp. These often occur in aging but otherwise healthy people; former president Ronald Reagan had one, and survived unaffected. In these sporadic cases, too, taking the polyps out before they invade the wall of the intestine prevents cancer. Otherwise, the disease becomes rapidly fatal.

One might begin by trying to find the gene for the more common, sporadic, form of colon cancer, the Ronald Reagan variety. Others did—among them Mark Skolnick, another genetic researcher at the University of Utah.

But Ray White felt it was a more sensible strategy to look at families where the contribution of heredity was overwhelming, "to pursue the cancer genes that you're sure of," he told me, "which were retinoblastoma [a rare eye cancer, of which more will come later] and polyposis. Those, although rare, are palpable, they're real. Whereas the others might be there, or might not. I thought the important thing to do first was to look where there's some light."

Not much light, however: There was no good clue about where among the 23 pairs of chromosomes to look for the purported gene for polyposis. He did have the families, "inherited" from Eldon Gardner (whose last graduate student became White's first postdoctoral fellow). Responsibility for the families passed to Mark Leppert, who set about finding other families with the disease, as geneticists in the lab began testing hundreds of DNA probes, at random, looking for a linkage to one of them.

"Slot machines," Leppert said to me. "You put your quarter in and you pull the arm, and you just keep running them." In the same field of research, looking for a marker linked to Huntington's disease, James Gusella in Boston had found a linkage among the first ten markers he tried. After testing two hundred markers, Leppert saw no signs of a polyposis gene. "It's chance," he said. "Unless you have the right clue."

Then the right clue appeared. In January 1986, a report in the *American Journal of Medical Genetics* described a man who had multiple hereditary abnormalities, including mental retardation and familial polyposis, and was missing a piece of the shorter arm of chromosome 5. Leppert didn't see the report until midwinter.

Immediately, the lab shifted its attention to markers on the short arm of chromosome 5, to search for a linkage to polyposis there. At the time, there was only one good probe to be had in that region of the chromosome. It was an anonymous restriction-enzyme fragment called C11p11, a piece of DNA of no evident importance, which had been donated to the lab by the English scientist who created it. It showed a clear pattern in the DNA from Eldon Gardner's kindred and others with familial polyposis; the marker was inherited with the disease.

By July, the researchers had sufficient evidence to link what they called "familial polyposis—Gardner syndrome," and therefore the tendency to precancerous polyps of the colon, to that marker on chromosome 5. Leppert and White wrote a research paper, and mailed it off to the journal *Science*. It arrived in their editorial offices on July 23, 1987.

Three weeks later, a report by the English geneticist Walter F. Bodmer and co-workers appeared in a British journal, *Nature*, linking polyposis to the very same marker. So it goes. The British team had noticed the report about the mentally retarded patient sooner, they had begun working sooner, they found the linkage sooner, and they published more quickly.

"That's just the way it worked out," Leppert says. "I mean, I had run two hundred probes through. If I had run C11p11 through a year earlier, it wouldn't have been the same month. If one of us and not the other had been aware of the paper [about the patient with the deletion], there would have been only one paper. It's sort of the way it goes."

Science published White and Leppert's paper anyway. It was, at least, a confirmation of an important finding. A gene behind polyposis had been pinned down.

To understand what has happened afterward in colon cancer research, we need to return to the matter of retinoblastoma, the rare eye cancer. It was the subject of Ray White's first noteworthy contribution to the understanding of cancer.

In the early 1980s, when he was looking for an appropriate way to apply the newest tools of genetic research to human DNA, retinoblastoma seemed one of the best places to start. White's lab had been the first to break up human DNA specifically in search of genetic polymorphisms. By then he already possessed scores of DNA fragments that revealed such polymorphisms, many of them linked to known regions of certain chromosomes.

At the time, a substantial body of evidence already suggested that, in the case of retinoblastoma, a single gene was working alone to cause cancer. The evidence pointed toward a particular small region of a particular chromosome, number 13.

In a few children with retinoblastoma, that region of the chromosome appeared to be missing when geneticists analyzed their karyotypes by spreading the chromosomes on a microscope slide, staining them to create bands so they could tell them apart, and counting them. Compared to a search for genetic markers, karyotypes were crude. And the one thing Ray White knew how to do, perhaps better than anyone else at the time, was to peer closely into small regions of particular chromosomes, using genetic markers.

Then an even better reason arose for White to focus the lab's attention on retinoblastoma. In 1983, William Benedict, Robert Sparkes, and their co-workers at the University of California in Los Angeles reported a discovery that pointed even more clearly to chromosome 13 as the site of the retinoblastoma gene. They found a close linkage between the inheritance of retinoblastoma and levels of a certain protein, an enzyme called esterase D. The gene for that protein had already been found, in the same region of chromosome 13 that had sometimes vanished in retinoblastoma patients.

In several families with the rare eye cancer, they reported, children who developed the tumor produced only half the levels of esterase D that their healthy parents did. Furthermore, the protein was polymorphic; it occurred in two different forms, a difference that could be detected by electrophoresis. In a particular family prone to retinoblastoma, the cancer always coincided with one or the other variant of the protein.

The cancer could coincide with either variant, a fact that deflected attention from esterase itself as being directly involved in the cancer. But it was an important clue to the genetics, because losing a gene somewhere near the esterase gene clearly predisposed a person to retinoblastoma.

Even more intriguing was one girl whose normal retinal cells did not show any evidence of a deletion in the relevant region of chromosome 13, and produced the normal levels of esterase. But when the UCLA team examined cells taken from her tumor, rather than from normal retinal cells, one member of chromosome 13 was missing from the karyotype. And the tumor cells did not produce any detectable levels of the enzyme. The child herself did not have obvious genetic damage inside chromosome 13. Her tumor did.

"We maintain that there is only one retinoblastoma gene, and that this gene represents a prototype of a class of human cancer genes characterized by a loss of genetic information at the . . . tumor level," Benedict and his co-workers wrote in the journal *Science*.

■   ■   ■   ■

The hottest news in cancer research at the time was the discovery of oncogenes, which are apparently normal human genetic sequences that appear to be able to trigger cancerous transformation if they are introduced into the "wrong" location in the genome of cells in the laboratory. In human beings, the same thing might happen as a result of a genetic accident, or the oncogene could be introduced by a virus. In either case, the insertion of an oncogene at some locations seemed to have the effect of leaving a molecular switch for cell growth frozen in the "on" position.

It was puzzling, trying to reconcile these oncogenes with the retinoblastoma discoveries. Oncogenes are like intruders who cause mayhem by actively disrupting normal operations. What happened in retinoblastoma, apparently, was the complete disappearance of some genetic material. How could genes that were *missing* cause cancer?

An explanation had existed for more than a decade. In the early 1970s, the pediatrician Alfred Knudson had proposed that the problem in retinoblastoma was not the inheritance of a gene that caused cancer, but the loss of genes that protected against it. Knudson merely envisioned such anticancer genes; at the time, there was no way to find them.

Children with retinoblastoma might have suffered damage to, or even lost completely, genes that normally act to prevent cancer, he suggested. Looking at entire families, this would appear to be a dominant trait (like susceptibility to polyposis), but on another level it would work quite differently.

Knudson's challenging theory required medical researchers to shift focus, away from human beings who can inherit genetic errors, and down to single cells. As well as inheriting faulty genes from a parent, after all, someone can sustain damage to his own genes during his lifetime. Single cells can just as easily pass mutations to their own descendants. Thus a lineage of dividing cells—but not necessarily the descendants of the whole human being, unless those cells are germ cells—confers a flawed genetic message. Like all mutations, many of these flaws might be harmless or trivial. But some could cause cancer.

Knudson began to think about lineages of cells because he knew that children with the familial form of retinoblastoma developed their cancer at a much earlier age than those in whom it seemed to be an isolated event. It occurred to him that there might be two different ways to develop retinoblastoma, which would explain the distinction.

A susceptibility to retinoblastoma might be a hereditary trait within a family, if a single damaged copy of a gene, caused by a mutation in a sperm or egg, descended from generation to generation. Someone with such a gene would arrive on earth with one strike against him, when it came to retinoblastoma, but because he still possessed one normal anticancer gene (inherited from the other parent), the cells in his retina would remain normal unless a second genetic accident occurred to him sometime after birth. That damage could disable his one normal gene, which had been acting until then to restrain his retinal cells from proliferating.

On the other hand, the disease might sometimes arise spontaneously—both protective genes altered or eliminated, in successive accidents during one lifetime. Either way, a retinal cell without the protection of either anticancer gene would become a retinoblastoma, a tumor.

Unorthodox and complicated, the theory was also compelling. It meshed with growing evidence from animal research that many kinds of cancer required at least two events in order to develop. As cancer researchers say, a cell had to suffer two "hits"—two different carcinogenic chemicals, perhaps, or a hereditary predisposition from birth and then exposure to an environmental trigger during life—if it was to begin multiplying out of control, to become a cancer.

The discoveries from Benedict's lab clearly pointed to an anticancer gene involved in retinoblastoma. How to find it? "Ideally," the geneticists wrote, "it would be useful to have a genetic marker that is more polymorphic than esterase, and is also closely linked to the retinoblastoma locus."

Well, White knew how to look for such a marker, and so did his co-worker Webster Cavenee. Cavenee spent more than a year creating genetic probes from the implicated region of chromosome 13, searching for restriction-enzyme fragments of DNA that might be more closely linked to the cancer than the esterase gene was, and then testing them on cell lines from retinoblastoma families. He did not bother much about cell lines from the blood, which would reveal only the chromosomes the children had inherited. He focused on cells from the retina, searching for cases where genetic markers were present in normal cells but absent from the tumor cells.

"Lo and behold, in the first five that we looked at," White recalls, "there were something like two that showed chromosome loss, and one that showed mitotic exchange," in which an uneven distribution of chromosome fragments during cell division confers two identical fragments from one member of a chromosome pair on a daughter cell, rather than one copy from each member of the pair. As a result, some daughter cells might inherit two deletions rather than one.

The genetic marker studies confirmed that regions of chromosome 13 were actually missing from the chromosome, with a certainty that visual analysis of human chromosomes could not assure. They also provided markers closer to the gene than esterase D.

With Benedict and several others, White and Cavenee announced the discovery of new genetic markers for retinoblastoma in 1984. Having closely cornered the retinoblastoma gene, they made it possible for other researchers to run it to ground.

Three years later, after isolating the normal DNA sequences that are deleted from that region of chromosome 13 in retinoblastoma, researchers at the Harvard Medical School and the Whitehead Institute for Biomedical Research in Cambridge, Massachusetts, reported discovering a specific gene whose absence predisposes someone to retinoblastoma. Later, geneticists in California showed that inserting the newfound retinoblastoma gene into cells in lab dishes could prevent them from growing in the abnormal way oncogenes would provoke them to grow. The term *anticancer gene* was an abstraction no longer.

Ultimately the retinoblastoma gene was implicated in other human malignancies: breast cancer, small-cell lung cancer, and osteosarcoma, a bone cancer. Another anticancer gene was found on a different chromosome. These genes hold the key to the control of cell growth, and thus, in part, to the control of cancer.

The problem in familial polyposis was subtly different from that of retinoblastoma. What was inherited was not a susceptibility to cancer per se, but a susceptibility to tremendous numbers of

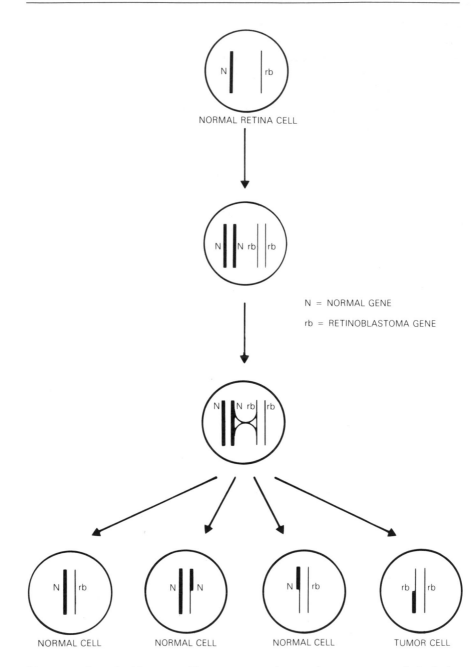

Mutations in retinoblastoma. The mutant and normal genes may switch during cell division, endowing one cell with two copies of the mutant gene although the original cell had one normal copy. Other accidents, such as deletion, may have the same effect: a cell inherits no normal gene that would protect against cancer.

benign growths that eventually, over a number of years, develop into cancer. Just because a linkage had been discovered did not imply necessarily that a gene on chromosome 5 caused, or perhaps protected against, cancer. It might merely be involved in the growth of polyps.

This was the question White asked Yusuke Nakamura to address when Nakamura first arrived in the lab. What was the presumptive gene doing? Nakamura began by testing whether there was something like an oncogene involved in polyposis: He tried to put DNA fragments from polyp cells into colon cells in tissue culture, to see whether something would provoke abnormal growth.

For technical reasons, the experiments proved futile. For one thing, Nakamura's assistant was never able to grow normal colon cells well in tissue culture. After that, he began to try to track down the elusive gene itself.

He spent about six months setting up a library of new probes. The term *library* is misleading; a DNA library is made of DNA itself, not of words written about DNA or symbols representing it. A DNA library consists of many dishes containing colonies of bacteria, stored in an incubator. Each dish contains a different set of DNA fragments, which are stored inside the bacteria, as part of plasmids, which multiply as the bacteria do. Libraries of clones are the raw material for genetic markers; they are the archives through which geneticists search to find good probes for disease linkages.

The main problem with most existing probes in the relevant area of chromosome 5, with the exception of C11p11, was that they were not polymorphic enough: Far too many people in a family would inherit the same marker type. So it was impossible to detect a difference between normal chromosomes and those carrying the faulty gene in a particular family.

The first "arbitrary" human DNA polymorphism ever discovered (the first one not related to a known gene) provided a clue to a way around the problem. That polymorphism, that variation in the fragments created by a restriction enzyme, had been discovered by one of White's postdoctoral fellows at the University of Massachusetts in 1980. It represented not a qualitative difference—a difference in the genetic sequence of the fragments—but one of quantity. People tended to inherit multiple copies of the fragment, but the number they inherited would vary. Some people inherited fifteen times as many copies as other people, and there were many gradations in between.

White was persuaded that there were other "tandem repeats" in the genome, and he set Nakamura to looking for them. They represented a powerful new way to find additional markers, because the degree of variability between people was greater.

Most variations caused by random mutations, an alteration in DNA subunits, existed in only two forms. If an adenine subunit was sometimes altered to a guanine in the DNA, for instance, there could be only two variants of the DNA sequence, and a pedigree would show only two types of chromosomes. This is the case, for instance, with sickle-cell anemia. But with tandem repeats, a father might inherit four copies on one chromosome, and six on another. His wife might inherit three copies on one chromosome, and five on another. There might be as many as eight different genetic types among their children—and other families could show entirely different types.

By analogy, it would be much more difficult to verify a child's parentage by his blood type, if all we had to go on were the Rh types, positive and negative. Adding to them the ABO types makes it much easier to follow a line of descent.

The use of tandem-repeat probes had been championed by an English geneticist, Alec Jeffreys, who applied them to a most celebrated purpose: helping to apprehend the man who raped and murdered two fifteen-year-old girls in the quiet rural county of Leicestershire. Jeffreys was looking for the most distinctive kind of differences; he had set out to test every man in the county and identify one, based on unique DNA patterns from his semen. Nakamura had a different need; he wanted to look for tandem-repeat patterns that were *shared* between people with the same disorder. So his technique was slightly different, but the principle was the same: He needed a very high degree of polymorphism.

By the end of 1988, searching for tandem repeat probes in the region of chromosome 5 linked to polyposis, Nakamura had identified six additional markers for the disease. They were highly polymorphic, and also more closely linked to the disease than C11p11.

Testing the new markers in the cell lines from polyposis patients, he resolved an old question about those different kinds of polyposis. Genetically, none of them seemed different from Gardner's syndrome. Every form of polyposis that appeared distinctive to doctors showed a linkage to the chromosome 5 markers.

A gene on chromosome 5, therefore, confers the tendency to develop colon polyps, but not necessarily the subsequent cancer. "By analogy to retinoblastoma, the simplest way to understand the polyposis gene would be that the way its [malignant nature] is revealed is to lose the normal copy," White said. "It was our expectation, based on this simple analogy, that that was what we'd see." It was not, however, what they found.

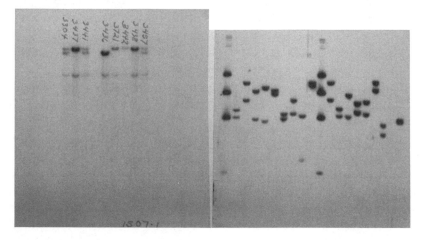

Two X-ray films of hybridization blots, demonstrating the difference between an ordinary polymorphism and a tandem repeat. Blot at left shows at most three different bands, representing at most three different fragment sizes in any lane. Blot at right is far more polymorphic; any person may have many different fragment sizes, and the patterns are therefore more distinctive. Courtesy of Yusuke Nakamura.

In Baltimore, Bert Vogelstein and his co-workers at the Oncology Center of Johns Hopkins University had been looking at

the problem from another level: the cellular. Vogelstein was attracted by the multistage nature of colon cancer. It seemed to
him that each stage might represent a different gene, and researchers in his lab had been using DNA probes (some of them
provided by White's lab) to look for differences between chromosomes taken from normal colon tissue, from polyps, and from
colon cancer cells, carcinomas. His subjects were tumors, not
patients.

One of the things the Baltimore team found early on was
that over three-quarters of the colon carcinomas had lost part of
chromosome 17. They also tested polyps; almost none of them
had deletions in that chromosome. As a courtesy (because he
knew White had a particular interest in polyposis), Vogelstein
called to tell him that his probes had helped to implicate chromosome 17 in a later stage of the colon cancer process. They
began to phone each other quite regularly to share ideas, and in
due course they decided to collaborate.

In his own lab, Vogelstein told me, "we started out working
on colorectal cancer because we can see these stages of progression"—from normal colon to polyp to cancer. Nakamura had
been testing genetic markers on the polyposis families. The
studies complemented each other, and eventually merged.

The result was an exhaustive search for evidence of genetic
abnormalities among the chromosomes of intestinal cells from
various stages of the cancer process in both kinds of cases, the
sporadic and the familial. Collaborating with White's group and
with that of biochemist Johannes Bos in the Netherlands, the
Baltimore researchers could see signs of a chain of molecular
mishaps.

As colon cells evolved toward cancer, they tended to accumulate several specific genetic changes. About four times out of
ten, in isolated polyps taken from patients with nonfamilial colon cancer (those who had not had polyposis beforehand) part
of the long arm of chromosome 5 was missing, the region Nakamura was working on. In the full-blown cancer cells, other chromosomal regions were often missing: many had lost part of the
short arm of chromosome 18, and 60 percent had a mutation in
an oncogene known as *ras.*

By the time cells had progressed to the carcinoma stage—
were capable of invading below the surface of the colon and of

spreading to other organs—nearly 80 percent had lost genetic sequences on both chromosomes 17 and 18. Although the changes were not invariant from one person to another, they did tend to correlate with specific stages of progress toward cancer.

"Our results support a model in which accumulated alterations affecting at least one dominantly acting oncogene [ras] and several tumor-suppressor genes are responsible for the development of colorectal cancer," the scientists concluded in the *New England Journal of Medicine*. It must take decades to accumulate all those mutations in four different chromosomes, assuming they are all necessary to nudge a normal colon cell from polyps to colon cancer. The odds against it must be phenomenal. When you think about it, it's obvious why colon cancer is most common after age fifty, and somewhat remarkable that it's common at all.

Subsequent studies in Vogelstein's lab (taking particular advantage of some of Yusuke Nakamura's probes) have suggested that the accumulation of genetic damage continues even after the polyps have turned to cancer. They revealed that patients with more deletions in their tumor cells are significantly more likely to develop distant metastases or recurrences, and also more likely to die of cancer.

Differences of this sort might explain why cancer is much more virulent in some people than in others, Vogelstein pointed out, and why some people respond better to anticancer drugs. In some research centers, this kind of DNA analysis is already being tested as a way to plan cancer treatments.

Because the most common deletions in colon tumors are regions of chromosomes 17 and 18, Vogelstein decided to focus on these chromosomes. One feature of chromosome 17 was especially interesting: At the dead center the deleted region was the site of a known oncogene, called p53.

That was something of a conundrum. Why would the loss of a gene that caused cancer drive a cell to evolve into cancer? It was the Knudson question all over again, but they couldn't use Knudson's explanation. The p53 gene was known to be an oncogene—one of the active intruders—not an anticancer gene whose absence would explain the growth of a cancer.

Vogelstein's team tried every biochemical strategy they could think of to explain the question. They tested the p53 gene in the member of the chromosome pair that did not have a deletion, to see if that p53 was somehow abnormal. But they saw no indication that it was. For a long time, they decided p53 must be uninvolved. They forgot about it.

Then, during the following year, a development from another lab prompted them to look closer. It appeared that the previous studies that had labeled p53 as an oncogene were in error. Unknowingly, those researchers had been dealing with a mutant of p53. In their follow-up study, the normal p53 gene behaved like an anticancer gene; it blocked the transformation of cells to cancer.

The Knudson hypothesis might hold after all in this case. Colon cells might progress toward cancer by losing an anticancer gene on chromosome 17—assuming the remaining gene on the other member of the chromosome pair was also abnormal.

The suggestion is fascinating, but was there really some mistake in the p53 region of colon cancer cells? The rest of us can enjoy speculating; people like Bert Vogelstein get paid for turning suggestions into facts. Vogelstein decided his only recourse was to forget trying to devise a clever, elegant approach and simply to apply brute force to the problem: working on normal cells, he would determine the complete DNA sequence in the p53 region; then he'd compare it with the sequence in his cancer cells.

Finding complete DNA sequences is "doable," he told me. "But it's just a pain in the neck. You only want to do it if you have to, in order to prove or disprove a point. Anyway, we did it, and to our initial surprise and delight, we found a mutation."

In colon cancer cells, where one member of a chromosome 17 pair had a deletion, the corresponding segment of the other member of the chromosome pair had a minor abnormality, a mutation. Such a colon cancer cell had no normal counterpart of p53. No anti-oncogene.

His later studies revealed mutations of p53 in tumors taken from many other human tissues: lung, breast, brain, ovary, cervix, and bladder. Close study of the p53 protein shows that most of the mutations occur in regions where it is chemically identical in humans, birds, and amphibians. By implication, these regions must have a very important function yet to be discerned.

The chromosome 17 story comes to a neat resolution, but some puzzling facts remain about the findings from Baltimore. Not all of the tumors showed all of the chromosome changes Vogelstein has discovered. Less than half of the polyps showed a loss of the genetic marker Nakamura had found on chromosome 5. Even more curious, none of the polyps from polyposis patients had lost that particular stretch of DNA.

Vogelstein believes that the undiscovered gene on chromosome 5 may be important in those other cases, too, but that it does not arise from an accidental deletion. Perhaps a hereditary mutation explains it instead. In the mutated form, he speculates, the gene may not be able to suppress cell growth. But without the ability to study the gene itself, speculations aren't worth much.

Unlike Vogelstein, Yusuke Nakamura has no p53, no known ''candidate gene'' on chromosome 5 to point him straight at the mutation. Until one emerges, or until he finds himself close enough to the gene that it's practical to analyze the DNA sequence subunit by subunit, we can't know what's wrong on chromosome 5, what kind of assaults provoke a normal colon cell to begin making too many descendants of itself.

Let's stand back now, away from the deep interior of tumors and cells, and consider matters from the public health perspective, the realm of warning signs of cancer and expert advice to eat more dietary fiber and less fat. What difference does all this close scrutiny of DNA from tumor cells make to you and me, and to those 150,000 Americans in any year who learn for the first time that they have colon cancer?

For an answer, I consulted a medical doctor who specializes in colon cancer, and one who has a special perspective on the question at hand. He is Dr. Randall Burt, the principal gastroenterologist who finds new patients for both major genetic studies of colon cancer at the University of Utah, Ray White's and Mark Skolnick's.

If the Nakamuras and Vogelsteins could present us with the identity of every gene involved along the progression to colon cancer, and with an easy way to test for it, he explained, those matters of warning signs and anticancer diets could be pinpointed to the 150,000 people in particular. Such a test could identify those people who most urgently need to take fat and fiber seriously, because they already have some mutations in their cancer genes. The rest of us might take oat bran or leave it.

More important, it might resolve a problem of fatal embarrassment, he went on. The public image of colon cancer is distasteful: People think that detecting it involves an embarrassing and terribly uncomfortable medical exam, and that if the doctor finds something they'll have to have surgery and wind up wearing a pouch outside the abdomen for body wastes, which has to be emptied.

Both perceptions are wrong, and often fatally so. New methods of surgery have been developed, and almost no one has to have a colostomy—to wear that little pouch. Not even one of the familial polyposis patients, who has to have his entire colon removed before he's old enough to vote, needs to have a colostomy. The solution may be draconian, as White told his students, but nobody needs to know about it. After the surgery is healed, it may be necessary to use the bathroom four or five times a day rather than once or twice, but that's all.

The exams have changed remarkably, too. In Eldon Gardner's day, a rigid colonoscope nearly a yard long had to be inserted up through the rectum. It's unpleasant even to think about; no one went through the procedure casually.

"For most people, it's the indignity of the colon exams that prevent its general application," Burt told me. "These old rigid proctoscope exams were quite uncomfortable. People had to be in a position where their bottom was in the air and they weren't draped. Whereas now people are well draped, they're lying on their side, the flexible instrument is much more comfortable, it doesn't take any longer, and it examines about three times more. Most people who've had it don't mind coming back for more screening."

If everyone over the age of fifty would be tested in this way, he added, more polyps would be discovered earlier; even the cancers would be discovered more quickly; and the total cure rate would be 75 percent or 80 percent, not 50–50 as it is today.

If there were a blood test for susceptibility to the cancer, he went on (really dreaming now) the cure rate could approach 100 percent. People who knew they were at risk would probably be more willing to have colon exams. For people who were not, it might not even be necessary.

Not long after Dr. Burt joined the faculty of the University of Utah Health Sciences Center, he sought out Mark Skolnick to explore whether they could merge their interests in colon cancer. It was sometime in 1982.

From studies of large families with hereditary cancers, Skolnick had discovered that colon cancer—the very common type, not Gardner's polyposis—appeared to cluster in certain families. He had even found a large family with several cases of colon cancer in close relatives, although they never developed polyposis. But the nature of the heredity was spotty, neither clearly dominant nor clearly recessive.

"He was trying to figure out how he could show whether or not this family really did inherit colon cancer, rather than just have it happen on the basis of common environmental factors, or some other reason," Burt recalled. "So I suggested we look at the rest of the family." Maybe if they looked at the healthy people they would find a clear hereditary pattern for *polyps*, if not for cancer.

Luckily for Burt and Skolnick, not to mention the subjects of the study, the flexible proctosigmoidoscope had recently come on the market. This made it considerably easier to persuade healthy members of a family to be examined for colon polyps. Eventually they collected 34 kindreds, and tested 670 individuals.

"The first thing we found," Burt said, "was that the family members had about two and a half times as many of these precancerous polyps as the spouses, who supposedly would have been exposed to the same environment." With the help of statistician Lisa Cannon-Albright, they reanalyzed the pedigree, counting both intestinal abnormalities—polyps and cancer—as signs of a hereditary condition.

Now the pedigrees fit a standard Mendelian model: the most likely explanation was that in these families a dominant trait conferred the tendency to colon polyps and, later, colon cancer, on some individuals. The upshot of the study was astounding. About a third of the people in the total population, their analysis suggested, might carry such a gene for this predisposition.

"The genetic predisposition is common, not rare," Skolnick explained to me. "It is responsible for a majority, not a minority, of colon cancer." The result appeared in the *New England Journal of Medicine* in 1988.

"You know, with pedigree analysis, it's still only a very likely hypothesis," Burt pointed out to me. "The ultimate proof of our hypothesis that it's genetic rests with finding a genetic marker. Statistically and mathematically it's very likely, but in fact, it's not proven until we find the markers." When that happened, he went on, not only would it prove the genetic hypothesis; it would allow doctors to identify the only individuals who really need to be tested for polyps.

A few months after the article appeared, I asked Skolnick whether he was working to find the marker. "We are trying to map the gene," he replied. "It is a long, slow process. We have no positive results to report yet."

However, by that time White and Skolnick were trying to collaborate to answer an obvious question: Do families prone to the common, sporadic colon cancer, if they share a predisposition to polyps, also show a linkage to the same markers on chromosome 5 that Nakamura has found in the familial polyposis families? Is Nakamura's unfound gene, perhaps, the one that every third person possesses, if you accept Skolnick's estimate?

During a phone conversation, White told me that such a study was considerably more "challenging" than the earlier analysis of polyposis families. For a cancer susceptibility gene as common as the one Burt and Skolnick predicted—possessed by one in three individuals—it was very likely that any new person who married into a family might have the chromosome 5 gene linked to a different variant of the marker from the one the rest of the family had. The polymorphism pattern might shift very rapidly, almost erratically, within single families. How could they sort it out?

"We're getting near the ultimate test of whether linkage can tell you something important," he said. "We're really depending on the isolation of that gene next year." With the gene in hand, they could look directly for a mutation responsible for polyps. They might test any individual for a predisposition to polyps, rather than entire families—just as your blood type can be determined without drawing blood from your parents.

Why, I asked White, do you say "you're depending on the isolation of the gene next year"?

"Because a year ago I said next year, and we still don't have it. So next year." Over the phone, I could hear that he was smiling. "Hurry up, Nakamura."

Once you've come close to a gene, as Yusuke Nakamura has, how can you tell where and what it is? Genetic linkages are useful mostly to indicate the rough location of a gene on a certain chromosome, and to pare down the region of interest. They never deliver the gene itself.

The classic way to narrow the search—and the one so arduous that everyone tries to avoid it—is *walking* the chromosome, as Lap-Chee Tsui did to pin down the cystic fibrosis gene. In walking, you try to find new markers that overlap a marker you already know about but lie closer to the suspected disease gene. You can tell if they do, because your new marker will be inherited separately from the disease gene less often, over the course of generations, than the old marker is.

You will probably have to do this over and over again. Walking may require testing hundreds of overlapping markers, inching slowly along the chromosome, in blood cells from hundreds of people.

An enhancement of this strategy is gene *jumping*, devised by Francis Collins at the University of Michigan, who used it to help find the cystic fibrosis gene. In jumping, you create a very large DNA fragment that overlaps your original marker, and then you make it a circle, by joining the two ends of the fragment in the same way you would close a necklace (for genetic engineers, this maneuver is simple). Once you have "captured" the distant end, farthest from your original marker, you can use it to determine whether the disease gene is still farther down the chromosome, or within the large fragment you have created. From there you can either jump again, with a new large fragment, or start to walk in the right direction inside the existing one.

Eventually, with luck and perseverance, you may find a genetic sequence so close to your suspected disease gene that it is never, ever separated from the disease by recombination. They absolutely always travel together down a pedigree, your marker and the presence of the disease. You're right on top of it now— or so the genetic studies say—but you may yet need to search through thousands or even millions of DNA bases to find the sequence that actually codes for the protein that has gone awry in the disease you want to solve. Now what?

Ideally, like Vogelstein with p53, you already know about a candidate gene within the region; it seems like a logical suspect to test. It may always be a red herring, of course. In the case of manic depression, the gene for a brain chemical known as tyrosine hydroxylase was located very close to the site pinpointed in the Amish study. It turned out to be irrelevant (as, apparently, was the linkage itself in that case).

Another way to search out the protein is to try a *zoo blot*: you use your probes to test blood cells from a variety of other species. Any DNA sequence that is conserved in a wide variety of species is probably of fundamental importance. In the case of muscular dystrophy, zoo blots showed that one region researchers had identified in human DNA also occurred in cows, mice, hamsters, and chickens. That helped to clinch the discovery of a faulty muscle protein.

Still another common approach is to look for what geneticists call *message*—genetic sequences complementary to part of your region—outside the cell nucleus in tissues affected by the disease you are studying. This is a strong clue, a sign that those cells are translating some section of your DNA region into protein, which would imply that the DNA sequence in question has functional meaning to the cell. In the case of cystic fibrosis, after an exhaustive search, scientists in Toronto found message for the cystic fibrosis gene inside sweat gland cells.

If you can translate the message according to the genetic code, locate the corresponding protein inside healthy cells and either fail to detect it or detect a mutant version in diseased cells, you have—like the cystic fibrosis team—made a major advance, and won the gratitude of thousands of sick people. Now all that remains is to figure out how to correct the problem you have identified. Back to the bench.

8

REVELATIONS

Many landmarks on the linkage map of human genes exist because of individual families: Cajuns, American blacks, Amish people and Mennonites, American Indians, Ashkenazi Jews, Scottish families, Icelandic families, Mediterranean families, and countless other families of mixed origins. Thousands of these families have contributed time, effort, and certainly blood to the identification of one spot on the map.

One group stands apart, because it has contributed more significantly to the gene map than any other set of kindreds, not just to certain points on the map but to the creation of the linkage map as a whole. These are the families of Utah.

What made Utah critical to the gene map was not merely that many of its families traced back to the same Mormon ancestors genetically; it was that so many of the citizens continued to share a philosophical heritage from those ancestors. The original Mormons saw genealogies and large families as matters of religious importance—and in the process created families of great value to science.

The history of human linkage mapping traces back to a genetics seminar at the Alta ski resort in Utah in April 1978. At the seminar, a graduate student stood up and began to describe his studies.

If the molecular genetics of humanity is now having its day, this was a few moments before dawn. The first study to link a human disease to a DNA fragment created with a restriction enzyme (the sickle-cell anemia linkage mentioned in Chapter 3) was still six months away from appearing in print.

The graduate student had been studying a family prone to hemochromatosis, a rare and sometimes hereditary abnormality of iron metabolism. He related how, in the family in question, he and his adviser, Mark Skolnick, were finding a genetic linkage between the disease and certain HLA types, the tissue-typing categories used to match organ donors and transplant patients.

He proposed that this linkage resolved a long-standing dispute about whether the disease was caused by a dominant gene or a fairly common recessive one. Until the HLA linkage, either explanation had seemed equally likely. The study had tilted the odds dramatically; Skolnick saw it as exciting proof of the power of genetic linkage to resolve opposing and complex explanations for the heredity of diseases.

The graduate student presented this first at a closed meeting, a private retreat for the genetics department at the University of Utah. As sometimes happens at these affairs, the participants burst into an unrestrained argument. It was the kind of dispute that still divides those geneticists who tinker with molecules from those who tinker with statistics.

Skolnick, in the statistical camp by training, was doing what he always did: looking for associations that would explain a pattern of inheritance, making rational assumptions based on the existing evidence about a disease. Unable to mate his study subjects at will (because he was studying human beings), Skolnick took it for granted that he had to put up with some uncertainties about any individual's genes. The best he could expect to find was the likelihood that a certain person had a certain gene. He had to speak probabilities. And he was proud of his new result.

The molecular biologists in the audience, by contrast, were used to using research methods that give binary answers about the presence of DNA sequences; yes or no. To them, what Skolnick's student was saying sounded like humbug, and they said so. They told Skolnick he didn't know up from down.

Two prominent geneticists from outside the university happened to be present as external advisers. David Botstein of the Massachusetts Institute of Technology and Ronald Davis of Stanford University were probably the only two people in the room who understood both sides of the argument.

Botstein had just finished mapping a gene in yeast, using restriction enzymes to make genetic markers in DNA. Davis had done the same with fruit flies. Botstein rose to his feet and inserted himself into the verbal melee. He came to the defense of Skolnick.

He had tried a different approach, but not an invalid one, Botstein argued. At the present time, all you could use for this is HLA, he pointed out to them, because in humans only HLA is polymorphic enough to use as a linkage marker—that, and blood groups, and a couple of other markers. But if you had polymorphisms all over the human genome, you could go anywhere, and get genetic variations, and find linkages to diseases. It would be normal to do that.

Botstein looked at Davis. Davis looked at Botstein. Botstein sat down. At that moment, only the two of them understood: quite likely other markers *did* exist, everywhere, all over the human genome. Restriction enzymes picked up all sorts of variations in the DNA of yeast and fruit flies. Why not humans?

Some time later the British geneticist Walter Bodmer made passing reference in a scientific article to the possibility of making a human genetic map using restriction enzymes. But it was unquestionably Botstein who picked up the idea and pushed it along.

His first step was to talk to Skolnick, who had almost no idea at the time what a restriction fragment was or how to blot DNA onto a filter. But over the course of a dinner conversation he did come to understand the brainstorm that had struck Botstein and Davis, and to share their excitement.

Looking back, Mark Skolnick prefers a slightly different description from "brainstorm." "The flash was a prolonged roll of thunder," he says, "with ideas bouncing back and forth, between brains and in protracted time." There was one critical question to answer: Would there be enough polymorphism in the human genome to generate a map?

Mark Skolnick works at a small research park affiliated with the University of Utah Medical Center. His lab is at the eastern edge of Salt Lake City, where the foothills of the Wasatch range, the western edge of the Rockies, rise abruptly. Crowding against them are the research park, a century-old army post called Fort Douglas, and the brand-new slabs and towers of the main medical center.

The major reason for Skolnick's presence in Salt Lake City is up a canyon to the south of the city. At the base of a massive cliff are the portals of another institution, which has strong but explicitly unofficial links to the university medical center. It is the Granite Mountain Records Vault of the Genealogical Society of Utah.

There, the descendants of the first Mormon settlers in the state have amassed by far the world's largest genealogical repository. The vault (which is sealed, climate-controlled, and protected against sabotage and against nuclear Armageddon) contains well over a million rolls of microfilm. They document genealogical information pertaining to at least a billion and a half souls from many nations all over the world—not merely Mormon ancestors. Almost certainly, some of your ancestors are in there.

The vault was the result of a Mormon doctrine that, according to one of the church's own pamphlets, "must occupy a high place on the list of 'unusual,' 'curious,' or 'peculiar' practices." The doctrine holds that the deceased can (indeed must) be baptized and rescued from damnation, long after death, through a sacrament performed by the living.

The need for posthumous baptism, based on a firm belief in bodily resurrection, justifies an untiring effort by armies of Mormon disciples to gather genealogical records of all kinds from parish records and government offices around the world, wherever they can. Among the most complete genealogies, not surprisingly, are those of the first Mormon pioneers who arrived in Utah.

The conservators of the Genealogical Society are quite happy to let people use the data base to trace their own family trees.

For several years, they also gave Mark Skolnick access to the data base, to trace Mormon families in particular, in order to study the heredity of common cancers.

Eventually Skolnick ascertained and computerized some 170,000 families, descending from the original Utah pioneers, to make his own genealogical data base. Within a few years, it contained entries for about a million people of Mormon descent.

During that time Skolnick began to link his Mormon genealogy by name and birth date to records in the Utah Cancer Registry. "By linking a population-based tumor registry to a large genealogy, we can study familial clusters of cancers and evaluate the genetic components of each site or appropriate subdivision," he wrote in 1977. (He also studied a family with hemochromatosis.)

Today the Mormons remain very fruitful ground for medical genetics. There is now a special office at the University of Utah dedicated to ensuring that people who use the resource Skolnick tapped do so wisely and ethically. Utah researchers are careful now to point out that they do not target the Mormons specifically, but search out their kindreds from patients at the Utah medical center the way geneticists everywhere do. The fact remains that most large families ascertained in Utah have strong if not exclusive roots in the Mormon community. More than two-thirds of Utahans still refer to themselves as Mormons.

In his book about human kindreds, Alex Shoumatoff calls today's Mormons "one of the nation's healthiest, most fertile, most cohesive, and most prosperous enclaves." For a number of reasons, they are perfect for genetic studies.

The Church of Jesus Christ of Latter-Day Saints, popularly known as Mormonism, sprang up in 1830, based on a revelation that led a young man named Joseph Smith to announce that he had discovered a new gospel. The first Mormon settlers arrived in Utah in 1846, after a long trek westward. Like the Old Colony Mennonites, they migrated in search of peace. But they had been driven west by real violence at the hands of people who resented their political strength and, even more fervently, reviled one of their doctrines: polygamy.

With that exception, the Mormons were sober and clean-living by anyone's standards. They foreswore alcohol, tobacco, and caffeine. But, as the result of another revelation, Mormon leaders had adopted "plural marriage" to increase the number of Mormon saints.

The polygamy doctrine would tend to exert some founder effect on the current Mormon population, giving them a somewhat limited set of genes because certain ancestors had contributed their genes to a relatively high proportion of the community (the Mormon patriarch Brigham Young himself had 53 children and 301 grandchildren). But the doctrine was outlawed by Congress and then abandoned in 1890, and its effects must be somewhat attenuated after a century.

The major feature of the Utah population that attracts geneticists is its large, stable families. Utah fertility is higher than national averages in any category you could choose. According to research at Brigham Young University, Utahans begin having children earlier and at a faster pace than other Americans, and do so as long as they are fertile. Families are also more stable than the national average.

Like the Amish and the Mennonites, the Mormons also have a clean-living ethic that helps to eliminate environmental factors that contribute to disease. But they are not particularly unusual in the genetic sense, deriving from a diverse group of American converts of mainly English, German, and Scandinavian ancestry. This may make genetic studies of Mormon families more relevant to most Americans than those of Cajun, Amish, or Mennonite people.

The Mormons' mark on Utah is inescapable. The major landmark of Salt Lake City today is the great temple. The first Mormon settlers of Utah chose the site for the temple four days after their arrival.

Today the city stretches miles to the south and west of the temple. The blocks are set out on a due north-south axis in a grid that dates back to the first streets laid out by the Mormon settlers, based on a plan for the city of Zion that came to their founder, Joseph Smith, in another revelation. The streets were named for numbers and letters of the alphabet.

Today, many newer streets have descriptive names such as Mango Road and Mountain View Drive. But roughly half the street addresses in Salt Lake City are still as colorless as longitude and latitude markings. You might address a letter to 3000 South 500 West, Salt Lake City.

Up the hill at the medical center, stored in thousands of bacterial plates, are massive collections of equally colorless addresses in DNA, taken largely from the blood of Mormon volunteers. They are anonymous DNA fragments, created in order to study diseases but also, equally important, to become locations on the general map of the human genome. It is one legacy—perhaps the most important of all—that Joseph Smith would never have envisioned.

A few days after the seminar at the ski resort, Skolnick, Davis, and Botstein turned up together downhill in Salt Lake City, at the lab of geneticist Ray Gesteland in the University of Utah's biology department. They were bubbling with enthusiasm.

They had figured how many markers it would take to map the human genome, Gesteland recalls. "Their feeling was that it would take three or four markers [to prove the concept] and, geez, if we could get a hundred of these, it's all over.

"They had great enthusiasm. What wasn't really clear was who was willing to put himself on the line and do the work."

Neither Botstein nor Davis was prepared to halt his own research midstream and start searching for polymorphisms in human DNA. At the time, Skolnick didn't have the expertise. But within a few months, the problem was solved. Botstein got an unexpected call from an old acquaintance, volunteering for duty. It was Ray White.

White told Botstein he had learned about the new idea from his former graduate adviser at MIT, who had heard about it from Skolnick at a breast cancer conference in Washington. This was several years before White came to Utah. He had been watching his promising early career drift downward from fruit flies to blowflies and toward an oblivion of esoterica. But he had just the skills to do what Botstein needed.

"This was the neatest thing since sliced bread," White recalled, "and something I wanted very much to be part of. So my response was to call Botstein . . . tell him I heard about it, indicate my interest, and ask whether there was a role I might play."

"His response was quintessential Botstein: 'Wonderful! I'm so glad you called! I've been thinking about it, looking for somebody to pick up this project, to do this thing, and you were the second person on my list.' " Remembering this, White erupted into a high, loud belly laugh. "Which is also quintessential Botstein. It was not done on purpose, it was not done maliciously, and I've never asked who the first person was."

Ray White got turned on to molecular biology during college, in Oregon, in the 1960s. He was drawn to the molecular biologists, seduced by their enthusiasm and energy. "They stood head and shoulders above everyone else," he says.

After gaining a Ph.D. in the biology department at MIT (where he met David Botstein), White moved to Stanford University for postdoctoral training. There and then, scientists were learning how to clone pieces of DNA inside living bacteria.

Studying the DNA of fruit flies, Ray White began to develop an intricate method for picking DNA segments of interest out of a collection of different clones. Complications arose; he became enmeshed in trying to interpret what they meant, and whether they were even interesting. After a few years, he moved to the University of Massachusetts Medical School, where some other scientists had noticed the same phenomenon in blowflies.

"I would have been much better off to stay another year or two at Stanford," he told me much later. "I made a mistake. I'm very lucky that it was not to be a critical mistake."

White spent several years studying the DNA of fruit flies and blowflies. In the interim, a team of Harvard researchers demonstrated the same phenomenon White had found—intervening DNA sequences, apparently meaningless genetic material that interrupts genes—in the human hemoglobin gene.

"It became clear that intervening sequences were a big deal," White said, "but not for fruit flies! So it was clearly time to find something else to do. As luck would have it, I got on this gene mapping tack." He says it as if it was a lark. Maybe it was, but it made his name.

White knew the technology required to look for polymorphisms in human DNA using restriction enzymes. He was also more than willing to commit his resources and energy to the effort.

"At the time, it was a hypothesis," he said. "Is there a human polymorphism we can detect with restriction enzymes? Yes, there ought to be. The next question was would there be enough of them to actually find one, just doing some sort of reasonable searching scheme?" Quickly he set his lab assistants to breaking up human DNA clones with restriction enzymes.

If they did find a DNA polymorphism in humans, would it be useful? The intervening sequences just discovered by Philip Leder's team at Harvard hinted that the elements of the human genome might be completely fluid. They suggested that in some way irrelevant material might be inserted willy-nilly around the genome.

Geneticists were used to thinking of genes as fixed, lined up like houses on a street, always at the same addresses in different people's chromosomes. But perhaps it would turn out that human DNA had a tendency to flit about, within an individual's genome, during his lifetime. The human genome might be either too unstable or too invariant to map. Nobody knew.

With contributions from Botstein, Davis, and Skolnick, White applied for government funds to support a research project which could begin to answer these questions. Together, they all wrote an article, largely based on the grant proposal, describing their concept for human gene mapping. It appeared in the *American Journal of Human Genetics* in 1980.

As a mere hypothesis, not the report of a discovery, "Construction of a Genetic Linkage Map in Man Using Restriction-Fragment-Length Polymorphisms" was a very unusual contribution to the literature of molecular biology. The new DNA technology, the authors pointed out, provided "a theoretically possible way to define an arbitrarily large number" of polymorphic fragments. At the time of writing, in other words, they had absolutely no idea whether it would work.

The article went on to define restriction-fragment-length polymorphisms (or RFLPs, as they called them), to explain how they might be used as genetic markers along the human chromosomes and, ultimately, given the right degree of polymorphism and the right kinds of human pedigrees, might form the basis of a high-resolution map of the entire human genome. "We are initiating our search for RFLPs in large Mormon pedigrees with genetic defects," the authors wrote.

Briefly, without much comment, the article went on to point out some possible implications: improved prenatal diagnosis, the opportunity to determine in advance one's own susceptibility to inherited disease, and a potential for a better understanding of human evolution. "The application of a set of probes for DNA polymorphism . . . ," wrote Botstein et al., "should provide a new horizon in human genetics."

The article has become a classic in molecular genetics, cited hundreds of times as a reference. By the time it appeared in print, White's postdoctoral fellow Arlene Wyman had already found one polymorphism in human restriction-enzyme fragments. The DNA had come from a number of people in Mormon families, and from some unrelated individuals. They named it for her: pAW101.

Shortly after the discovery of that first RFLP, White was recruited to the University of Utah, in an arrangement that involved the Howard Hughes Medical Institute. The institute was created by the famous multimillionaire in 1953 to support medical research, on the proceeds from his aircraft company, formed by breaking it away from the Hughes Tool Company on the very day the Medical Institute was founded. By 1981, the institute's budget was nearly $25 million a year.

Rather than setting up its own labs, the Hughes Institute pays the salaries and costs of scientists where they already are. A Hughes grant is like found money; you can enhance your situation immensely without necessarily disrupting it.

The institute was interested in genetics, but it was not especially committed to the University of Massachusetts Medical School, or to blowflies. When officials of the institute offered Ray White a Hughes grant, he told me, a relocation to Utah was part of the bargain.

It was an obvious place to continue to look for more RFLPs in large families. Besides, Mark Skolnick was already working there. At thirty-seven, "happy as a pig in mud," White relocated.

As a genetic statistician and a molecular biologist, Skolnick and White should have made a powerful team. But that's not how it worked out. Between the two labs, as one of Skolnick's close colleagues puts it, "there's been a history of up and down." At times, they could barely be civil to each other; at other times, they collaborated without friction. There may have been a personality clash. Part of the dispute involved who would oversee the gene-mapping work, and how. White, the molecular biologist, wound up with the job.

In any event, within a few years, the two geneticists were working separately in the same general field, but with a different emphasis. Although at the same university, they work in two different laboratories separated from each other by the Fort Douglas army base. Skolnick continued his studies of common human cancers (breast, colon, and skin), and began to hire staff who could learn the new techniques. White and his team began to generate new RFLP markers by the dozens, aiming to create a genetic map and also to resolve the heredity of diseases as the opportunity arose.

The idea of mapping the human genome by genetic linkage was not new, of course. It had been suggested years earlier by Victor McKusick. The difference was that now there was a practical way to do it.

"After I'd been here several years," White told me, "it became apparent to me that we knew enough about the system—how to find markers, how to generate genotypes, what kind of families to use, all this kind of stuff—we knew how to do this project. It was just a matter now of getting it done."

There was no question anymore that the human genes were located at fixed addresses. It soon became evident to everyone involved that making a map of them was far too much work for a single scientist or small group. Finding enough useful RFLPS would entail sorting through hundreds of DNA fragments, or thousands.

Shortly after the idea emerged, David Botstein approached an official of the National Institutes of Health, to propose that the government fund the mapping effort on a large scale. He was politely rebuffed: the institutes funded basic research into diseases, not the development of tools for a molecular genetics project.

Obviously someone did mount a major project, but the

money did not come from the deep pocket of the U.S. government. As it turned out, the map came to exist because of a private philanthropy (the Hughes Institute) and a private company.
The human genome project is based on a feasibility study that
the U.S. government got for free.

After being turned away by the NIH, Botstein suggested the
project to Collaborative Research, Inc., of Bedford, Massachusetts, one of the nation's oldest biotechnology companies, for
whom he and Ronald Davis both served as scientific advisers.
Other than Ray White's lab, it was probably the only operation
in the world prepared and equipped to begin scanning human
DNA for RFLPs.

Collaborative was nearing its twenty-fifth anniversary, and
operating in the red. Nonetheless, the company saw a reason to
take on human gene mapping. It might be able to use or sell all
those markers later, for medical diagnosis.

Initially, Collaborative tried to interest Wall Street investors
in the mapping project, but a downturn of the market in 1983
killed the initiative. Meanwhile, to demonstrate to investors its
own commitment, Collaborative had already begun to search for
new RFLPs, and got itself drawn into the search for the location
of the gene for cystic fibrosis. Ultimately the firm decided to back
a mapmaking project itself, on the prospect of earning its money
back someday, in DNA diagnosis.

White moved ahead on his own. He had decided that the
best way to fund human gene mapping was with a consortium
of private foundations that support hereditary disease research,
such as the Muscular Dystrophy Association and the Cystic Fibrosis Foundation. Collectively, he thought, they might be willing to pay for the creation of a genetic map that would eventually
benefit all of them.

One day in 1984, White presented these ideas to Hughes
officials and advisers who had come to Utah for a site visit, to
review his progress and determine his future funding. His audience included "a very first-rate group of people, frighteningly
first-rate," he recalls. Among them was Dr. Donald Frederickson, the former head of the National Institutes of Health, who
had just become president of the Hughes Institute.

White told his audience what the team was doing toward

building a gene map, why they couldn't do the job on the funding they had at that moment, and what he had in mind to accomplish with the coalition of charity groups. He finished, and fell silent.

Frederickson looked directly at him. "Well, you know," he began, "we've been thinking about this before we came, and actually the kind of effort it would take to do this through a coalition would be really a substantial outside diversion for what we think a Hughes investigator ought to be up to."

"Oh!" White coughed out, taken aback. "But you know, I've been discussing this with George [Cahill, a vice president of the institute] all along. I thought he understood and agreed to this."

"Well, we did approve of the explanation," Frederickson said. "But now we've considered it and indeed, we feel it's inappropriate." There was a rather long silence. "On the other hand, we think it's rather a good idea, so we'd like to fund it."

"He persuaded me on the spot," White says. His funding doubled, in an instant.

White has found himself quoted in the press in the context of two major competitions, both times pitted against Collaborative Research. His quotes tend to be put in opposition to those of the head of Collaborative's gene-mapping effort, Helen Donis-Keller.

Early in the race to find the cystic fibrosis gene, Collaborative had asked White to join its effort, and he agreed, but the collaboration never really came into being. His lab was happy to send its probes to any other researcher after the probe's existence had been published and sometimes before, but when the agreement with Collaborative failed, White stopped sending them his probes. Collaborative, a business with future profits to worry about, considered its own probes proprietary. At this point, White and Collaborative began to compete to find the cystic fibrosis linkage. Ultimately, they published a linkage on chromosome 7 at virtually the same time, and then White dropped out of the race, to pursue other genes.

The other friction arose over the first human linkage map. To judge from the published accounts, Collaborative won a contest to create the first map. But there are problems with that view of events. Ray White would say he wasn't really in such a race, that it would have been the wrong race. He was the tortoise not the hare, publishing maps chromosome by chromosome, as he felt he had pinned enough markers to a chromosome to make a useful map.

In October 1987, after spending five years and nearly $12 million, a team of thirty-three geneticists affiliated with Collaborative Research (among them David Botstein) published genetic linkage maps for all the human chromosomes at once, in the journal *Cell*. The lead author was Helen Donis-Keller.

"We had ideal circumstances that were not possible in an academic setting," she recalls. "What you really needed, especially in those early days, was a dedicated group of people who would stay with a project for a long period of time. [We had] a very large component of professional-technical-level workers. The industrial setting was really ideal."

With great fanfare, Collaborative unveiled the map at a meeting of the American Society of Human Genetics in San Diego. Donis-Keller announced that the map, with more than four hundred randomly spaced markers, gave any scientist a 95 percent chance of finding the general location of a disease gene on any chromosome.

The markers were spaced an average 10 centimorgans apart (a *centimorgan* is a measure of genetic linkage roughly equivalent to a physical distance of a million DNA bases). Some chromosomes were especially dense with markers, notably chromosome 7, where the cystic fibrosis gene was located. A press release announcing the map estimated the size of the market for prenatal DNA diagnosis based on the map at "several million dollars a year in the United States."

Ray White had at least as many RFLP markers as Collaborative did. In fact, some of their markers had come from his lab. In Ray White's view, the Collaborative publication was premature, and he said so to reporters who quickly called him for reaction to the announcement.

"It is not what we believe should properly be called a map," he told a reporter for the journal *Science*. He said he would never have dreamed of publishing a map with such "significant gaps." He added, "There is this feeling of having been burned. . . ."

Donis-Keller's retort came in the same article: What right does Ray White have to decide what's a proper gene map?

White was not quoted in reply, but he gave me an answer much later, privately. The right was not merely his, he argued; it belonged to all the members of the Centre d' Etude du Polymorphisme Humain (CEPH), an international consortium based in Paris, which had been organized expressly to create a universal gene map.

Both White's team and Collaborative were members of CEPH by the time of the *Cell* publication, and White felt the company was not playing by CEPH's rules, as he saw them. White had donated his best cell lines to CEPH, and thereby indirectly to Collaborative, which could draw the DNA out of the consortium's banks and test it itself. By leaping into the press with its own map, White argued, Collaborative had been unfair to the entire world of gene mappers, not just to Ray White and company.

There have been several very visible races in human genetics lately, searching for genetic linkages to cystic fibrosis, to Huntington's disease, to bipolar disorder, and there are doubtless other examples that have not gone public. Those researchers who have shared their probes and cell lines freely have had people tell them they are crazy, and scientists in both academia and private companies have been secretive about their results until and even well after publication.

Such behavior is hardly unique to molecular genetics, but the field seems to have particularly good prospects for reporters looking for controversy. As James Watson has pointed out, in the effort to create the human gene map, "competitive races dominate what should be a noble point in human history." In his own book about the discovery of the DNA double helix, Watson wrote vividly about what he called "the contradictory pulls of ambition and the sense of fair play." This is nowhere more obvious than in the history of human gene mapping.

It riles Ray White and a lot of other scientists that the press has chosen to focus on the heavy competition in this field. After all, White complained to me, probes do get shared, linkages are made, and genes do get discovered. The process works; why not focus on that?

Nonetheless, in a relaxed moment, he admitted that the

competition is not all bad. "After all," he said, "there's nothing like a good tennis match to sharpen your teeth." It's a startling metaphor, but you know what he means.

Given human minds and human DNA, the idea of restriction mapping of the human genome was probably inevitable. Given the human character, it could have happened much more slowly than it actually did. Beyond the private company (Collaborative) and a private philanthropy (Hughes) are the voluntary contributions of a French artist and a retired French scientist, which have enabled the mapping to proceed smoothly, and, in the end, with some elegance. The history of CEPH is a delightful counterpoint to all the intrigue in molecular genetics.

In 1965, the French immunologist Jean Dausset was one of several people in his field who, like Botstein and White two decades later, realized that he was working on a problem too big to solve alone. Dausset's own problem (which he later won the Nobel Prize for solving) was how to sort out the HLA transplantation types in the blood cells of human beings. Fifteen years before he won the Nobel Prize, he was one of the major figures involved in organizing a series of regular international workshops, where a researcher could share blood serum samples and analyze how his blood typings in Paris compared with those in, say, Virginia.

"It was the first time in the history of biology that researchers worked together and shared results" on such a scale, Dausset says. At the time, about a dozen labs were involved; today the number is more than three hundred.

In 1978, an admirer willed Dausset her complete collection of modern paintings, intending him to devote the proceeds to science. She died a year later. Dausset retired from science, won his Nobel Prize, and then turned his mind to the human genome, which was at least as urgent as tissue typing and, if anything, more complex. He decided to use the money to do for genetics what the HLA workshops had done for immunology.

At first, Dausset intended to collect a set of cell lines from large French families, and distribute them worldwide from his new Center for the Study of Human Polymorphism. Then, at a scientific meeting in Miami, Dausset met Ray White and learned about his Utah pedigrees. They were far larger than anything Dausset could hope to amass in France.

By that time White had stored up cell lines from dozens of Mormon families of three generations each, families that had the best characteristics to use as reference pedigrees for gene mapping. He knew that, as he puts it, "there would be a premium in encouraging other people to run their markers on our families," but he had no organized way to do so.

After the meeting, two associates of Dausset's hustled Ray White into a taxicab and, on the way to a restaurant, suggested with straight faces that he hand his best cell lines over to Dausset. In return, they offered, White could become a founding member of the repository.

White agreed, and twenty-five of the forty-odd families in CEPH's collection of cell lines are from his collection. CEPH's only reason for being is to make a human gene map by coordinating linkage studies from all over the world.

CEPH distributes the DNA to any researcher who can use it. In return, the researcher must send back the precise genetic details that emerge from subsequent research on those DNA samples, and any new probes. Most of the cell lines come from Utah, but there are many others, including one set from an Amish pedigree.

Some of the genetic analysis of these cell lines comes from Collaborative, some from White, the rest from many other labs. Way off in Paris, CEPH allows scientists to rise above their squabbles. By the time a map comes out of CEPH, it is almost anonymous.

A year after Collaborative's map appeared in the journal *Cell*, CEPH's data were already sufficient to prepare a map twice as dense, at the 5-centimorgan level, for human chromosome 10. It was based on data from some thirty research groups. Other chromosomes would follow, with predictable regularity.

"Research workers are strong in competition," Dausset told me, choosing his words carefully in a second language, "but they are rational, and they understand what is in the interests of science. We all work in one world, on the same genetic material."

With the CEPH data "it should be virtually impossible for any disease gene to escape," White said at a seminar in late

1988. "That's rapidly becoming true for every chromosome."
Ten years after a way to map the human genome suddenly
sprang into someone's mind, the map was nearing completion.

Even as it is being charted, the map of all the human genes is
useful. Someone like Martine Jaworski can dream up a molecular explanation for some human oddity, borrow or make probes
that recognize some DNA segment she suspects to be involved,
and look for a linkage in a suitable population. Or, like Kidd
and Gerhard and hundreds of others, a researcher can use
"brute force," testing probes from all the chromosomes at random in hopes that a linkage will pop out. The linkage moves
research forward toward a solution for that disease.

Meanwhile, the results of the search, linkages and nonlink-
ages, go back into the collected information about the human
genome. The information either strengthens or weakens the evidence that a certain gene or DNA fragment is actually in its
assigned spot on the genetic map.

The popular image of a chromosome is suddenly very out
of date. Most people who have a mental image of a gene probably think of a spot on a fuzzy, microscopic line with dark and
light stripes, the chromosome pictured in a high school textbook. This is chromosome 7, stained to create such an image:

Human chromosome number 7, chemically stained to create light and dark bands.
Human chromosomes are banded in this manner for prenatal genetic testing.
Courtesy of Dr. Helen L. Drwinga of the Coriell Institute for Medical Research.

That's not a gene map; it's an earlier, much more crude version, a low-resolution photograph made using stains that stripe the chromosomes.

What is being created by Jaworski and Nepom, Bronya Keats, Daniela Gerhard, and others too numerous to name is something much more useful. It's a dense, linear list of alphanumeric labels or names representing different DNA fragments—either DNA probes created by restriction enzymes, or genes associated with diseases, or small regions of DNA known to create proteins that occur in several normal variant forms, such as hemoglobin.

On the genetic map, all of these markers are keyed as closely as possible to the region of the chromosome where they are located, and are lined up in order. This is also chromosome 7:

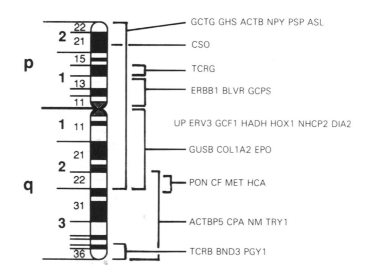

Linkage map of human chromosome 7. (From Victor A. McKusick, "The Morbid Anatomy of the Human Genome," *Medicine* 66, January 1987, p. 26; copyright by Williams & Wilkins, 1987. With permission.)

This list of unpronounceable labels, the map of genetic markers, is the result of many studies of DNA patterns from the blood of many large families. What has made the map is count-less pieces of information about how often or how rarely two markers are segregated—de-linked, physically separated—during the course of reproduction. If they are never or almost never broken apart and exchanged by a genetic recombination be-tween two chromosomal arms, they're probably quite close to-gether. The distances between markers on the map are generated by LOD scores, which can be used to assign the probable loca-tion of any marker relative to any other marker nearby.

Most human minds can handle only the information nec-essary to analyze this relationship for one pair of markers at a time. As long as there were relatively few known human poly-morphisms, such as blood and HLA types, this represented no problem. But once the prospect of almost unlimited polymor-phisms in the DNA arose, it presented a serious dilemma: they could never be mapped in a reasonable time or with reasonable accuracy unless it was possible to compare results from many different linkages at once.

Even the existence of computers did not solve the problem. On the best of supercomputers existing in the early 1980s, using the traditional methods of linkage analysis, it would have taken prohibitive amounts of computer time—on the order of years—to resolve the number of markers that the geneticists were churning out with their restriction enzymes.

The solution, independently arrived at by scientists in White's lab and at the Whitehead Institute in Cambridge, Mas-sachusetts, and Collaborative Research, was a new type of com-puter program that can perform *multipoint analysis*. The new strategy, now available in fairly simple programs that work on personal computers, analyzes various possible combinations of linkages from a number of different markers. In one of the first reported applications used by CEPH, sixteen markers were mapped simultaneously during a computer analysis that took a mere nine seconds to run.

As it exists today, the human gene map consists of hundreds of DNA markers, all lined up in an order based on an international research consensus. Without the restriction fragments that link up to hereditary diseases—without those hundreds of poly-morphisms discovered in human DNA, starting with probe Arlene Wyman 101 (pAW101)—there could never have been a map much more detailed than the fuzzy Xs in a biology textbook.

Ironically, Collaborative's linkage map published in *Cell* was its farewell to the gene-mapping effort. By that time, the company had effectively stepped out of the mapping effort. A quick market for DNA probes in disease diagnosis had failed to materialize.

"The impact on us is quite disastrous," Collaborative's president, Orrie Friedman, told me. "The financial community doesn't understand all this. . . . When we talk to the financial people about the genome project their eyes glaze over. They think we're idiots to have spent all this money."

Financial experts understand the creation of new medications, he explained; beyond that, they can't fathom what market the genome project would serve. Collaborative shifted its focus to using DNA analysis for paternity testing and forensics. It may also, ultimately, make a great deal of money from people doing genetic mapping if it can draw on a license granted by Stanford University, which filed for a patent on the concept of mapping using restriction fragments. The patent has yet to be granted.

"We believe we're sitting on a powder keg" with DNA diagnosis, Friedman continued. "We think this will turn out to be a major coup . . . if we all live long enough." Several months later, the cystic fibrosis gene Collaborative set out to locate was found at other laboratories. The company would have to buy a license to market a DNA test for the most common genetic disease, but Friedman still sounded bullish. "All the indications are that there will be such a demand for the test—and there are very few people with the ability to apply it—that there will be plenty of work to go around," he said.

Years after both White and Botstein wished it would, the U.S. government finally began to commit funds specifically to human gene mapping. President George Bush requested $156 million for the project for fiscal year 1991. The funding of the effort is anticipated to increase throughout the rest of the century, coming to a total of about $3 billion, including contributions from private sources like the Hughes Institute.

However, Hughes is also ceding most of its role in gene mapping to the federal government. It still provides funds to CEPH, but it curtailed its support of a human genome data base at Yale University, expecting the general public to take up the obligation, through the government.

Hughes also declined to renew the five-year grant that paid Ray White to develop new RFLP markers. But White's search for new probes continues, now paid for by the National Institutes of Health under genome project money. (Hughes is still paying him to search for certain disease genes.)

Helen Donis-Keller, who has moved to Washington University in St. Louis, is focusing her research on an effort to find the genes behind behavioral disorders and certain cancers. In the process, she also continues to chart some parts of the gene map.

I asked Ray White what he anticipated for the next decade of human molecular genetics. "More genes will be cloned, more genes will be mapped," he replied. "Each of these is likely to be an extremely interesting story. The story will not be so much the mapping or cloning anymore, but the use of those genes to work out the molecular biology of those systems."

"There's yet another round," he went on, "which is in a sense the ultimate genetic challenge, and that's to sort out the more common human genetic diseases and traits." Or, as he and three others pointed out years ago in their now-classic entry to the *American Journal of Human Genetics,* the application of a set of probes for DNA polymorphism should provide a new horizon in human genetics.

For a while, it's exciting to piece together a treasure map. Of itself, though, the map is not the real treasure. You use it to find the treasure.

9

CASSANDRA
AND THE CREW

There are many treasure seekers in human genetics, on the trail of one gene or another, often after the explanation for a disease. There's also another sort of adventurer roaming about the genetic landscape: a cartographer, an explorer, someone who sets out to plot a map of the chromosomes purely for the exercise of mapmaking.

It is like sailing across uncharted ocean, says biophysicist Charles Cantor. You hope there's something interesting out there. You hope you don't come to the edge and fall off—or ply endless ocean forever.

The captains of gene mapping expeditions are at some risk, but in this fleet the crew members are in gravest danger of vanishing entirely. The postdoctoral fellows, or postdocs, who carry out the manual labor of gene mapping, are bound by an old aphorism: publish or perish.

The mere effort to make a map, though it requires skill, is not in itself a sign of originality worthy of publication. How many postdocs will vanish, professionally speaking, while charting a region of our gene map? We can never know.

First, there were details to attend to. Cassandra Smith, microbiologist, general manager, and den mother of a genetics laboratory at Columbia-Presbyterian Medical Center in New York, consulted her list.

"We've been checking the inventory," she told the fifteen young scientists and technicians assembled in the conference room. Glassware is disappearing from the lab. "If it isn't labeled, we can't get it back from the dishwashers. If you see something that isn't labeled CRC, please label it."

CRC stands for Charles Robert Cantor, who sat directly across from her at the head of the table, as he always did at lab meetings he was able to attend. Lately, that was not very often. He traveled so much that when an immigration officer at Kennedy Airport had recently asked him where he was arriving from, he drew a blank. Charles Cantor was very busy making designs, scientific and political, to map the human genome. He was also in great demand as a speaker.

Cassandra Smith was more or less in charge of the lab itself—especially in his absence, especially when it came to details. But the glassware was labeled CRC. Nearly half of the grant money was in her name, but he had the office with the window. Scientists elsewhere were likely to neglect Smith when they spoke about the lab, or else were ponderously obvious about not neglecting her. It was an unusual partnership.

Without interruption, Cassandra moved on down her list. (Everyone in the lab used first names.) Charles sat across from her, hands folded in his lap, smiling benignly. He's a small man, neat, mustachioed, wiry—the very image of a postal clerk, but witty. Cassandra is dark-haired and dark-eyed, intense. She dresses like an art director. Neither one dawdles often.

For some reason, the housekeeping list was unusually long at the lab meeting on August 21, 1988. In addition, she had two people to introduce. A computer consultant was visiting, to study the lab before setting up a new computer system. Cassandra also introduced a science writer doing research for a book on gene mapping, and asked everyone to be helpful.

"That's about it," Cassandra said, reaching the bottom of her housekeeping list. There was a moment of silence. She looked down the long table in the conference room.

Not all eighteen pairs of eyes looked back. As at most regular meetings, some people doodled on their pads, others stared off to the side. Only a few (usually the same few) joined in the conversations and paid close attention. Those were the people who always sat at the table, not in the chairs ringing the room, near the walls.

"Okay," Charles barked. "Shall we do some science?" He often started the real business of the lab meetings this way. Then everyone would wait to see who he was going to call on.

At most labs, it's arranged in advance who will give a progress report at the next meeting. Here, it was often a surprise.

Charles looked to a corner of the room, where a postdoc lounged in a chair along the wall, under a bookshelf.

The postdoc from Europe wants not to be named; things did not go well for him in the Cantor-Smith lab, and to identify him might jeopardize his future career. So think of him as Nikolai. "What's happening?" Charles asked Nikolai.

"Not much," Nikolai said from the corner. "I tried two experiments with PCR, and none worked."

PCR was the latest vogue among molecular geneticists. The acronym stands for polymerase chain reaction, which describes it: a chain reaction using a polymerase, an enzyme that makes copies of DNA. PCR was becoming pivotal to gene mapping, because of the speed with which it could multiply DNA fragments.

"Well, stand up and tell us about it," Charles persisted. Part of the purpose of lab meetings is to share results; another purpose is to train people to describe their work to others, whether or not the others already know something about it.

Nikolai ambled to the blackboard, and described briefly how he was trying to make multiple copies of a certain DNA strand using *Taq* polymerase, an enzyme extracted from a microbe that lives in hot springs. The enzyme survives the heat used in a PCR reaction, and saves a researcher from having to add a new batch of enzyme at every new cycle in the chain reaction.

"Somebody designed a machine to do this," he concluded. "We have one, and we are going to use it to do our work."

The latest versions of PCR could make a million copies of a specified small DNA sequence, such as a probe, in a matter of hours. Each cycle would copy the strands made in the last, creating four strands from two, and eight from four, and so on.

No molecular geneticist would explain a new procedure without scrawling an illustration. He will reach for any handy scrap of paper and sketch. See the picture of PCR on page 219.

Most labs that had PCR machines used them to amplify one or at most a few similar segments of genetic material at a time. It was being used most often for individual experiments, or to analyze specific known DNA segments in blood samples for crime work or for medical diagnosis.

Charles and Cassandra had grander ideas for their machine. They wanted to multiply scores of different probes already in use in the lab, and later, hundreds more.

Nikolai was simply trying to get the technique to work. As he described the time and temperature conditions he was using, he marked the numbers on the blackboard.

"Why did you choose those conditions?" Charles asked. Nikolai shrugged. "There needs to be a longer incubation. Call Francis Collins [of the University of Michigan, who regularly collaborated with Charles] and ask him about his conditions." He paused. "But what did you get?"

Nikolai shrugged again, with a tight smile. "Nothing."

"Well, let's see your data. Don't you have a photo you can show us?" Nikolai left the room for a moment, and everyone sat silent, waiting. (This usually happened; almost no one brought actual data to the lab meetings, so the speaker of the day was usually unequipped.)

He returned with a few Polaroid snapshots, rectangles of black scarred here and there with small white smudges. They were images of DNA fragments separated on a gel, which he had labeled with a fluorescent dye and photographed in the darkroom. Charles took one of the photos from Nikolai and stared at it.

"You've amplified little little garbage," he said. "This is crap. It looks totally heterogeneous." Charles handed it to Cassandra. "Do you have anything else?" Nikolai held out another Polaroid snapshot.

"The melting temperature, how'd you get it?"

Nikolai replied that he read the method in a published research report.

"Ah, but that's the paper we don't believe," Charles barked.

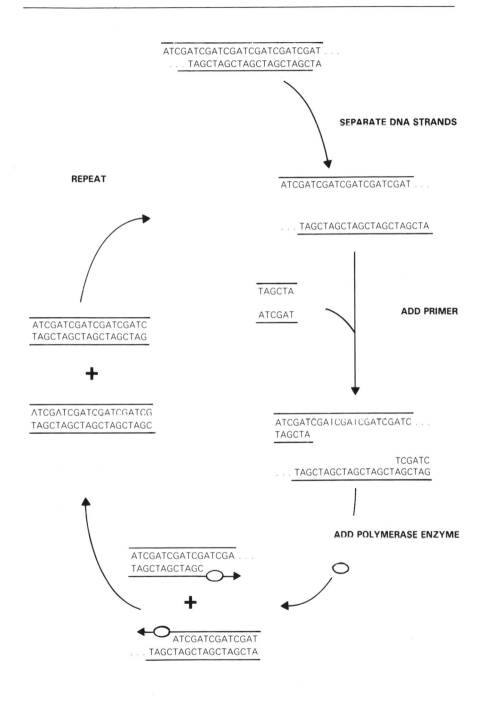

Polymerase chain reaction (PCR).

"Who?" asked someone. Cassandra mentioned a name, and the room fell silent.

"You've really amplified something," Charles mused. "I wonder what it is. . . . The oligonucleotide is too small. [Oligonucleotides are stretches of DNA.] Usually these things are meant to work on longer oligonucleotides. And the reaction temperature is way too high. So you'll have to use Klenow [another DNA-copying enzyme, which is not heat-stable]. You'll just use up a lot of Klenow, but that's okay. It's not nearly as expensive. You'll just have to add Klenow every time. That's what they did before they found *Taq* polymerase."

There was always a vivid contrast between what Charles Cantor could dream up and what a postdoc could accomplish in a given period of time. For four or five months, Nikolai tried out various conditions for doing PCR, and he slowly began to work out some of the conditions necessary to amplify the kind of DNA segments the lab would someday require in large quantities. Later, other scientists would take over the effort to get PCR working for the lab's purposes.

One of them told me that the technique is "generally magic in somebody's hands. Once they do it, they can do it reliably— but there are so many little nuances of what can go wrong.

"But Charles wants it now," he added. "Not six months from now."

It was a lab in transition. About a year later, the operation would disband and the principals, Charles and Cassandra, and whoever else was invited and willing to relocate, would move to California, to the Lawrence Berkeley Laboratory of the University of California. There, Cantor would oversee the portion of the American genome project that is being coordinated by the U.S. Department of Energy (which has supported genetic research ever since the inception of atomic energy, because radiation causes mutations).

Charles was directing an expedition to chart the physical map of the human genome, which is an entirely different atlas of human DNA from the one made by sorting out linkages between genetic markers and inherited disease. The physical map-makers pay little or no attention to what the DNA means.

They're interested in how it fits together. They chop chromosomes to bits, breaking them up at convenient spots, and then try to put the bits back together in order. It's like doing a jigsaw puzzle, or solving a code.

Genetic linkage studies can provide great detail about the genome, but only at close range. They can only estimate accurately the distances between nearly adjacent sites on the map, and assign genes to general locations. Linkage mappers are like city planners walking around the neighborhood with clipboards, noting down street names and house numbers. They can say roughly how far it is to a supermarket, but they can't give good directions to Albuquerque or Tampa.

The physical mappers are at an entirely different level, taking aerial photographs. Until quite recently, in fact, they were in orbit taking satellite images. Until the 1970s the best picture anyone had of the genome were whole chromosomes, visible through light microscopes—about as good as an earth map made from the moon. Then geneticists discovered how to stain chromosomes, to create distinct bands of light and dark color, like the image on page 210. That was the equivalent of satellite reconnaissance, with at most a thousand visible landmarks. Chromosome banding could identify the locations of chromosome segments within about 5 to 10 million DNA nucleotides.

Together, chromosome banding and genetic maps provided an aerial view of the continents, and also street maps of Fifth Avenue in New York, Wilshire Boulevard in Los Angeles, and Lake Shore Drive in Chicago. But there were no cross-country road maps, and no good estimates of the distances between cities.

What bridged the gap, and allowed geneticists to launch off on long-range mapping expeditions of the chromosomes, was a development called the pulsed-field gel (PFG), a variant of standard electrophoresis, which was invented by Charles Cantor and a co-worker in 1984. Just as ordinary electrophoresis separates proteins or small DNA fragments by differences in their size, PFG spreads pieces of DNA across a gel. But it can detect size differences between DNA fragments so large that they would all stay in a clump near the starting point on an ordinary gel, unmoved by the electric fields used in standard electrophoresis.

In PFG the orientation of the electric field is switched at specified intervals. This forces large DNA strands—between 20,000 and several million DNA subunits long—to bend and stretch, to snake their way through the same pores in an agarose gel that smaller molecules enter with ease. As they twist and turn, large DNA molecules travel differently, depending on their size. It's like a race between a pickup truck and a moving van. They may run neck-and-neck, but only until they come to a sharp turn.

It takes special enzymes to make such big DNA fragments. They're called "rare cutters," because they slice into a DNA molecule less often than most restriction enzymes do. The rare cutters have to find a specific eight-subunit sequence—not merely a particular succession of four DNA subunits—before they will nip a DNA strand. Just as the word *sequence* occurs less frequently here than the word *the*, specific eight-base sequences are less common than four-base sequences, so most of the stretches in between them are larger. One popular rare-cutter enzyme, called *NotI* (it's "not-one," rather than "not-eye"), creates fragments from a few hundred thousand to several million DNA bases in length.

Pulsed-field gels take longer to run—perhaps a week rather than half a day—and, for technical reasons, the lanes fan out. They can be more difficult to interpret than ordinary gels, and the technique is more elaborate and expensive than the gels that are the mainstay of the biological revolution.

The Cantor lab had twenty-three boxes, or pulsed-field gel apparatuses. The lab's major mission was to use the boxes for long-range mapping of various genetic regions. Cantor, and the physical mapping project in general, were on a rather tight schedule.

Just as it's easier to finish a jigsaw puzzle with large pieces, PFG gave physical mappers the first practical way to join large sections of the genetic landscape—to piece together aerial-size photographs of a suitable size to give a well-defined global context to all the local maps. The atlas isn't finished yet, but now it can be.

At Columbia, Charles was not merely Professor Cantor. He had an endowed position: Higgins Professor of Genetics and Development. He was also, of course, the developer of PFG, but by now he had advanced a long way from the laboratory bench.

Still, he loved designing and improving experiments, being a "fixer." As he put it himself, "I'm a very good consultant. I often have the ability when a project gets in trouble to listen for a couple of hours and get it out of trouble." The new DOE job was a chance to do this on a very grand scale.

"But I don't have time in some cases to get involved with the details," he went on. This, indubitably, was Cassandra's role.

Among many other duties, Charles was now engaged in staffing up the new gene-mapping lab in Berkeley. In California, the lab's budget was due to triple and its space would quadruple. Cantor told a reporter he expected his staff to grow rapidly to about a hundred scientists.

A year before the scheduled move, architects' drawings and photos of the new lab site hung on a bulletin board in the corridor outside the conference room. Here in New York, the lab was old and cluttered. The site was, well, New York. The photos from Berkeley showed wide lab benches, a view from the lab windows of dark shrubs and a broad valley.

But at the moment, despite this and other inducements, the only person in the New York laboratory who had committed to move to California was Cassandra. It was understood that their basic arrangement would continue there.

At Columbia, Cassandra Smith was an assistant professor in the departments of psychiatry and microbiology. Working in Cantor's department when PFG was developed, she used the technique to make the largest gene map of a one-celled organism. It was a map of the 4.7 million DNA bases of the bacterium *Escherichia coli,* which has been a standard model organism for biology research since 1940.

Smith and the co-workers paid under her own grants were busy trying to work out the three-dimensional properties of *E. coli*'s DNA. The smallest human chromosome, number 21, was only about ten times larger than the *E. coli* genome, and the same techniques should be broadly applicable to both. The *E. coli* work, her major interest, occupied about 20 percent of her time. The rest was spent on technology development and study of the human genome.

Charles and Cassandra worked together, and also separately. They had brainstorms together, and regularly squabbled about research procedures in front of the rest of the lab. They collaborated intensely and successfully. He was good at making experiments work, but to do that they needed data and samples, and she was good at providing both.

They took their collaboration home after work, and had been doing so for more than twelve years. "We have a tremendous amount of absolutely sheltered time," Charles said once. "Over dinner we can discuss science—which is most of our life— problems in the lab, or strategy.

"The fact that we come from different disciplines makes a tremendous difference," he added. "She's a biologist. I have no biology. I'm considered to be a rare example of a generalist." They also complemented each other in temperament, he went on. Charles called himself an optimist. Cassandra, he said, is a pessimist.

At length, released from interrogation, Nikolai returned to his seat at the outskirts of the lab meeting. Next Charles asked Aki-hiko Saito to come forward. Aki, as everyone called him, had joined the lab shortly after Nikolai began working on PCR. The mandates for both of them were the same; each must generate evidence of original research in order to further his career.

During two years in the Cantor lab, Nikolai was assigned one project after another, and never seemed to progress with any of them. He also generated a number of ideas of his own, but never seemed to find time to complete the experiments.

Aki's experience was quite different. He arrived in New York with orders to carry out one simple (or so he thought) assignment: to create a linking library of chromosome 21 from a collection of large fragments of the chromosome. A linking library is a chromosome broken in order to reassemble the pieces. His task was like breaking a necklace of multicolored beads and then having to figure out how to put them all back in the original order. But the pieces were invisible.

Aki came to New York from Japan with about 300 candidates in his library, fragments of DNA generated from human cells and cloned in bacterial cultures. Each set of cloned fragments might or might not be from chromosome 21, and might or might not be unique. He intended to spend a few months sorting all this out and return home, victorious and ready to publish the results.

Before coming to New York, Aki had been working in the Tokyo laboratory of Charles's friend, the Japanese geneticist Misao Ohki. A few months earlier, Charles ran into Ohki at a scientific meeting in Japan, and Ohki told him that he had begun to create a linking library of chromosome 21. Every fragment needed to be analyzed—to determine what part of the chromosome each clone came from, what sizes they all were, and which restriction-enzyme sites they contained, before trying to fit them all together.

The strategy, which Cassandra Smith pioneered, involves making two separate collections of DNA fragments. The first, the probes, are small DNA segments that surround the recognition site of a particular restriction enzyme. In the Cantor/Smith lab, where the goal was to make large DNA fragments and map them using the boxes, the favorite enzyme was the rare cutter NotI. Ohki had also made his library using NotI.

The first part of strategy was to cut chromosome 21 with a frequent cutter enzyme—something other than NotI—and then look for those fragments that had NotI sites in the middle. The second step was to break up a chromosome using NotI, and test it by hybridization with the probes. The probes can be used to identify any two fragments created by NotI that are adjacent on the intact chromosome, because they will overlap the contiguous ends of the two NotI fragments. (See illustration on page 226.)

Ohki had assigned this task to Akihiko Saito, who was a postdoc in his lab. It would be very helpful to use Charles's PFG equipment for the analysis, Ohki remarked when he saw Charles at the meeting. Ohki proposed to collaborate, to leave his linking library at the New York lab permanently, if Charles would allow Aki to analyze them there.

A medical doctor by training, Aki had gone on a sabbatical to Ohki's lab in order to learn molecular genetics so that he could use it later to study human disease. When he left his hometown of Niigata for Tokyo, he had left his apartment unoccupied. He meant to return after a year. After nine months, he had begun to discover some linking clones. Offered the chance to go to Cantor's lab, he jumped at it. Aki felt it would be relatively quick work to characterize the clones.

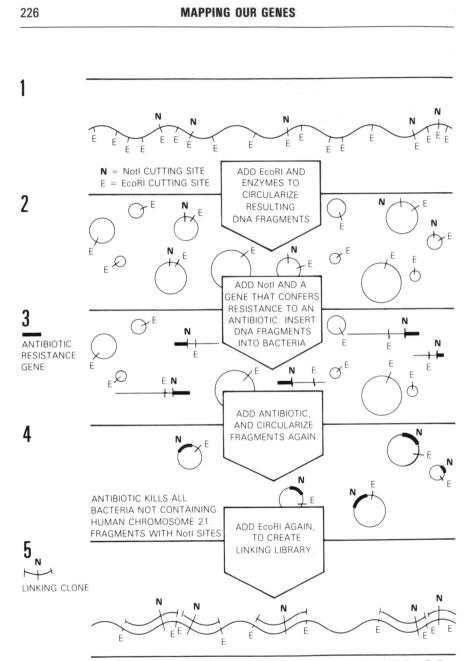

1

N = NotI CUTTING SITE
E = EcoRI CUTTING SITE

ADD EcoRI AND ENZYMES TO CIRCULARIZE RESULTING DNA FRAGMENTS.

2

ADD NotI AND A GENE THAT CONFERS RESISTANCE TO AN ANTIBIOTIC. INSERT DNA FRAGMENTS INTO BACTERIA.

3

ANTIBIOTIC RESISTANCE GENE

ADD ANTIBIOTIC, AND CIRCULARIZE FRAGMENTS AGAIN.

4

ANTIBIOTIC KILLS ALL BACTERIA NOT CONTAINING HUMAN CHROMOSOME 21. FRAGMENTS WITH NotI SITES

ADD EcoRI AGAIN, TO CREATE LINKING LIBRARY.

5

LINKING CLONE

LINKING CLONES WILL IDENTIFY NotI FRAGMENTS THAT ARE ADJACENT IN THE INTACT CHROMOSOME.

How to make a linking library.

Aki had come to New York expressly to learn which of the spoonfuls of liquid in his collection of Eppendorf tubes contained probes that would link up the strips of DNA broken off by NotI, and which did not. It was like trying to do a jigsaw puzzle, without a picture to go by, knowing that some pieces were missing and that the box also contained many extraneous and misleading pieces from other puzzles.

Aki meant to finish the job by September. Like postdocs everywhere, he needed a publication, a good credential that would help him get support for future research. Back in the internal medicine department at the University of Niigata, apparently, this is more an imperative than an incentive.

"I am in a serious situation," Aki told me bluntly, several months after we met. "If I cannot write a paper, I will be cut from my department."

Aki waited by the blackboard for Charles to speak. "How does the score card look?" Charles began. (The score card was their way of keeping track of progress on the linking clones.)

Aki chalked it onto the blackboard:

On 21	NotI cut	8
	uncut	4
Not on 21		6
???		18

At a glance, without regard to the breakdowns, it's obvious that the yield was, as yet, very low. Of the 300 fragments Aki had brought, he had tested only 36. Half of those were aligned with question marks. To call them "linking clones" was optimistic at best: the odds were very low that any two clones on the score card would actually identify regions of DNA that were adjacent on human chromosome 21.

In fact, as the score card showed, six of them weren't on 21 at all. Impurity—those extraneous and irrelevant pieces of the puzzle box—was a major problem.

At the outset, Aki had distilled his library from another library of chromosome 21 fragments obtained from Lawrence Livermore Laboratory in California. Scientists at Livermore had created this library using an automated process that sorts the different chromosomes. You could count on a certain degree of impurity; bits of other human chromosomes would slip through.

In characterizing his own library, Aki routinely included lanes on his gels to establish whether or not a clone actually derived from human chromosome 21. But more of that later.

Aki was testing his linking clones by seeing whether they matched samples of DNA extracted from a number of different cell types, and broken up. This technique was similar to what Alberto Severini had done with the DNA from Canadian Mennonites: Aki would separate the cellular DNA on gels using ordinary electrophoresis (not PFG). Then he would blot the DNA from gels onto nylon. Afterward, he would label each of his linking clones with radioactivity, and test it against the DNA on the nylon blots, to see which of the DNA from the cellular sources it bonded to.

"So we got five more?" Charles said. He did not sound impressed. Aki replied that he had been having some trouble extracting large DNA fragments from the gelatinous cubes in which they were stored. Besides the impurities, Aki had had difficulty getting the DNA he wanted in the first place: Some of his puzzle pieces appeared to be missing.

As a result of the extraction problem, some of his first blots probably had not contained enough DNA. Therefore, some of the earliest films had too little signal to interpret on the blot; the radioactive probes had found very little cellular DNA to bind to. Instead of sharp black-and-white ranks of X-ray signal, all he could see were gray shadows. His results had been getting better lately, he told Charles, so this might have been a temporary setback.

"So we have twenty percent NotI sites on 21," Charles went on. "That's very good." He paused. "In fact, I'm stunned by the fact that two thirds of the clones are cut. That would be terrific. It makes life so much easier."

He was referring to a third problem Aki had to deal with: some of the spots NotI ought to recognize in DNA are chemically masked in the native chromosome. The process of making linking clones unmasks them, and thus the probes might not correspond correctly to the cellular DNA. It was as if some of the tabs on his jigsaw puzzle pieces had been torn off. That

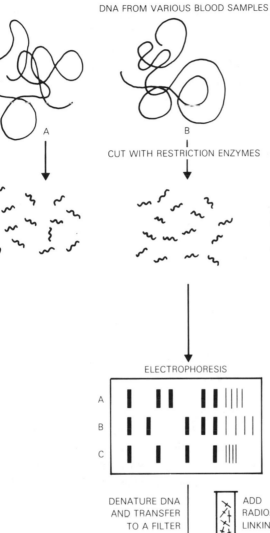
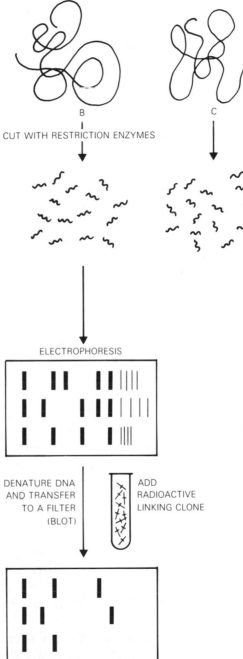

DNA hybridization.

explained the "Not cut" and "Cut" lines on the score card. Only the "cut" segments could match up properly.

Cassandra speculated that enzymes used in making the NotI library might preferentially pick out NotI sites that were unaltered in the intact chromosome, which would lead to fewer ambiguous results. "That would be nice," Charles responded. "That would be very nice."

"This stuff may be of immediate use," he added. "We may be able to start to map 21." He was too optimistic. That was only August.

Putting a person into print has a way of lionizing him. To be fair, there were six other postdocs in the lab, each of them working on projects of significance.

Among them was Jan-Fang Cheng, a native of Taiwan, who was quite close to cloning a *telomere*, the very end of a chromosome. For lack of a way to find the telomere, progress toward the Huntington's disease gene was stalled. Norman Doggett was working hard to map the region of the Huntington's gene in particular. Guy Condemine, the son of a family that owned a wine-producing château in France, was studying the structure of the *E. coli* genome, as was another American postdoc, Steve Ringquist.

Because of the impending move, the lab was so transient that Cassandra's personnel list was almost never up to date. It was becoming something of a scientific way station: some postdocs were preparing to leave, others were deciding whether or not to go to Berkeley, and staff members newly hired for the Berkeley lab were beginning to rotate through for a week or two, to get their bearings.

The corridors were narrow, lined with file cabinets on the diagonal and stray pieces of equipment. There were barely enough stools and chairs for all the people working there, but then they rarely spent much time sitting anyway. People moved busily in and out, passing each other wordlessly many times each day, making their own buffer solutions in one room, taking a gel to be read in another, taking samples to be spun in a centrifuge elsewhere. It was a lot like the scene behind the counter at McDonald's, except that no one stood and shouted orders, and it was okay to just sit and think—at least now and then.

■   ■   ■   ■

The 10:00 A.M. lab meeting on September 22 began late, as many of them did, because Cassandra was not ready at the starting time. It was Thursday, a bad day to be interrupted for a meeting. People wanted to get their PFG gels ready to run all weekend, so they could be used the following week.

When the word went around that it really was time for the meeting, people wandered into the library desultorily. Charles was away on business, in Italy.

"What's happening with the linking clones?" Cassandra asked Aki. "Any more problems? Let's see the score card."

"Not changed," he responded.

"Has changed."

"Has not changed."

"Why not?" she asked.

"Ahhhh—" It was often difficult to tell whether Aki was groping for the right words in English, or merely for an answer.

"What's the problem?" she persisted. She leaned forward and gazed steadily at Aki, waiting. Cassandra's dark eyebrows form a high arch, an inverted U that makes her eyes seem very large. The brow above them is often wrinkled as she waits for a reply. It has a way of making her look perpetually skeptical. It's easy to see why many people find Cassandra Smith intimidating.

"Some cell lines grow very slowly—than other cell lines," Aki replied at length.

"You've had contamination?" she asked, and there was a brief discussion about this.

"Aki," she went on afterward, "can you tell me what you've been doing? You must be doing something. You must have some films up or something. . . . I know you had several problems at the beginning."

"The score card has not changed," he repeated.

"Well, what have you been doing?" There was laughter. Aki, standing at the blackboard, smiled diffidently and waited for it to subside, then began to explain.

"First we have to . . . ah, answer some questions," he said. "First, whether our linking clones are really located on 21, whether the NotI sites are cleaved in the genome, and number three, whether our linking clones are really linking clones." He wrote the questions on the board, hacking at it with the chalk.

Aki's attack on chromosome 21 was as logically elegant as any cryptographer's assault on a newly discovered hieroglyph. Unlike a cryptographer, however, he already knew the code. But the tablet was badly fractured, and he had to reassemble it, by feel.

Aki had created electrophoresis gels that were thirteen lanes across. He had loaded the lanes with an array of DNA samples taken from different kinds of cells, which he had digested with different restriction enzymes in a highly logical sequence. (These were standard electrophoresis gels; it would not be time to use PFG until he was ready to try to link up the linking clones, once he knew which of his samples they were.)

The thirteen-lane gels were intended to answer a number of questions about an individual clone at one glance. Aki could indeed read the pattern taken from these gels almost at a glance. So could the technician who worked with him, Bill Michels. An X ray of one of his blots looked like this:

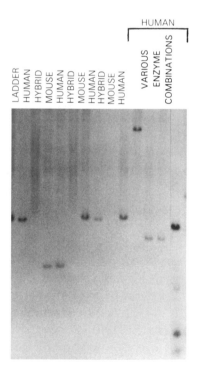

A hybridization blot of chromosome 21, made by Akihiko Saito. Courtesy of Charles Cantor.

At the far left of the blot, in the first lane, was a "ladder": a lane of broken DNA from a bacteriophage. These fragments of known size were used to measure the bands in the other lanes. To the right of the ladder, Aki had arranged the blots in ranks of corresponding lanes: three, three, three, and four.

Each of the three-lane sets contained DNA fragments extracted from three different cell types. In order, reading from left to right, the lanes represented pure human DNA, fragments of DNA from hybrid cells that had mouse chromosomes as well as human chromosome 21, and pure mouse DNA. The last four lanes at the right contained only pure human DNA fragments, broken with different restriction enzymes (other than NotI). This information would help much later, in mapping the chromosome.

Aki used the DNA in the first three lanes right of the ladder to determine whether a linking fragment had truly come from human chromosome 21. If a band appeared in the human DNA lane (lane 2) but not in the hybrid cell lane (lane 3), or if one appeared in the mouse lane (lane 4) but not the human lane (lane 2), then the radioactive probe had not come from chromosome 21. It deserved to be forgotten—at least for Aki's current experiment.

The second set of three lanes determined whether or not the NotI site contained in the fragment was chemically altered, cut, in intact human DNA. In the larger universe of science this kind of alteration is fascinating: it seems to indicate the starting point of genes, what a cell recognizes as a sort of chemical on-off switch.

But for Aki, the alteration was a minor irritation, a quirk to be accounted for because it would affect his attempt to correlate his library with the whole intact chromosome. If a band in lane 5 (representing total human DNA) stood higher than the corresponding band in lane 6 (mouse-chromosome 21 hybrid DNA), that meant the probe was detecting a larger fragment in the original human chromosome than it detected in the hybrid. Evidently, NotI could not recognize the site containing that probe's sequence in the intact human genome. That clone would not be useful for mapping.

The third set of three lanes contained DNA cut by MluI, another "rare cutter enzyme" that generated relatively large fragments useful for mapping with pulsed-field gels. If the radioactive bands in lanes 8 and 9 stood at a different location from the bands in the lanes immediately next to the ladder, Aki could conclude that the enzyme MluI had snipped the DNA somewhere inside the NotI fragment represented by his probe. (Lanes 4, 7, and 10, mouse DNA, should be empty.) This information would be useful for mapping, later.

Aki briefly described this whole scheme at the September 22 meeting. Then he went on to tell Cassandra about the problem that was keeping him in New York beyond his planned departure date. It was those question marks on the score card, relating to eighteen of the clones in his library. He had come up with some new explanations for the extraction problem he had discussed with Charles at the first meeting, the source of the shadowy, gray X-ray films.

"Now the problem is," he said, "many clones used for hybridization as probes didn't detect any [radioactive] signal. Maybe our probe DNA preparation was not so good. Or the other possibility is that the bands to be detected may be too large. First, our blotting was also not so good."

"You want to show us a nice film?" Cassandra asked. "Why don't you show us some data?"

Aki left the room and returned with a large, blue ring binder, full of his hybridization films. He leafed through it, and then clipped one to the light box in front of the blackboard.

"It may be that you're not detecting signal because of polymorphisms," Cassandra said. 'The first lane is human DNA, right, here?" She paused to think. "Let me ask the question that, of the ones on 21, did you see any polymorphism? Did they all detect the same signal, in human DNA and in the hybrid? Is it the same size fragment?"

"Yeah," said Aki. He clipped another, clearer film to the light box.

"Were there any that were different?"

"This band was a little lower than that," he said. "But the difference is always like that."

"Have you looked—these clones are from Livermore Labs, right?"

They were.

"Well, they have published what percent of the clones they think are from chromosome 21," Cassandra said. "Does your score card agree with theirs?"

"Ah, no," Aki said. "The purity of these clones is very low." Livermore maintained that at least 90 percent of its library contained fragments of chromosome 21.

One of the other postdocs made a comment, and there was laughter. Aki looked puzzled. "What Fang is saying," Cassandra said, "is that with the eighteen, with all those question marks, and the ones not on 21, you don't come out to anywhere near ninety percent. So the major problem is the eighteen."

"Eighteen," Aki echoed. He stood at the board, sighing now and again, waiting to be finished but not released. Cassandra suggested a few ways he might try to solve the problem, and finally brought the meeting to a close.

"We gotta get you to practice, Aki," she said as the laboratory staff rose to leave. "You're gonna give a talk." In less than a month, Aki was due to present the lab's linking clone data—such as it was—before a large audience of scientists at the American Society of Human Genetics meeting in New Orleans.

Aki smiled and looked down. "I do practice," he chuckled. "Practice at home." Somewhere nearby in their apartment, his porcelain-faced wife would listen intently as he addressed her on the subject of molecular biology (which she barely understood) in English (which she did not understand at all).

New York was daunting to her, and she had few Japanese acquaintances in the city. To buy Japanese groceries, she took a bus all the way across the George Washington Bridge into New Jersey. If only for her sake, Aki needed to finish his work quickly and get back to Japan.

Aki had extended his visa to December, because work was not progressing quickly. "The worst kind of data is no data," remarked Bill Michels, the technician who worked with him. Any one of many things can go wrong, he went on, but if something doesn't work you may never know which one it was.

I once heard Cassandra Smith joke about the ways to hide your secrets from competitors in the same field. One of the best ways, she said, is to have a postdoc who doesn't speak the language very well present your results at a scientific meeting. That way you can get credit for the work, and avoid confronting questions you don't want to answer.

Later she amended this statement. "In general, scientists vary the amount of information they are willing to share depending on who they are speaking with," she stated. "With a native speaker this may be very subtle, but in the case of a foreigner I guess it just comes out as bad English. In any event, you certainly do not have to use a non-native speaker to obscure if [you] want to do that." She chose Aki to speak at the genetics meeting in New Orleans, she had said, partly because no one in the lab would give a better talk about chromosome 21, and partly to give him the experience.

At the podium on October 14, facing a ballroom filled with geneticists from all over the United States and beyond, Aki fussed with the microphone draped around his neck, finally took it off, and held it in his hand. Then he leaned forward and launched into his address.

"I'd like to focus my talk on preparation and characterization of these NotI linking clones," he said. He briefly described the procedures, and flashed a few of the clearer examples of his multilane films on the projection screen, explaining what they showed. Then he put the score card on the screen.

Pulsed-field gel experiments, he had told the audience, showed that the average NotI fragment was about one million DNA subunits—1 megabase—in length. The whole chromosome is about 50 megabases long. Fifty or sixty NotI linking clones, he concluded, should be enough to map the chromosome.

"This is a tentative score of our NotI linking clones," he said, pointing to the slide. "These 10 and 5 clones have cleavable NotI sites on chromosome 21, in the genome of hybrid cells containing the human chromosome. . . . If the human chromosome 21 is supposed to consist of fifty NotI restriction fragments, we need to prepare about fifty NotI linking clones. That means, we have now thirty percent of the NotI linking clones necessary to complete restriction mapping." Preliminary results, one would have thought, but interesting.

There was the requisite applause, and then the chairman

asked if there were any questions. "Please ask me slowly," Aki said, leaning into the mike.

There were only two questions. How many probes are there altogether? There seem to be sixty or seventy candidates, he replied. How do you know that the candidate clones are different? someone asked. Because we have digested them with several restriction enzymes, he replied, and compared the patterns.

The score at the time was 15 clones, identified as being useful, unique, and part of human chromosome 21. Aki did not point out, of course, that he had come to the United States with 300 candidates to analyze, and fewer than 100 remained by the time he came to New Orleans.

The eighteen clones that gave no signal certainly were not worthy of discussion. Roadblocks and stumbling blocks are matters to be aired only at lab meetings, if at all—and certainly not in front of a national audience, which is why the public seldom hears about them.

The next Monday, he was back to the routine: flicking Eppendorfs over and over with his finger to dissolve DNA in minuscule drops of solution, rolling a glass rod across a plastic bag again and again to drive air bubbles out of the way of his blot after adding the probe solution, pouring buffer solution from yardlong cylinders into plastic boxes.

At various times he compared his routine to cooking, but one afternoon he came up with a different metaphor. "A farmer," he mused. "If he plant in the spring, maybe he will get a crop."

Everyone in the laboratory had an assigned section of "bench," the deep counter that runs at waist height under a shelf or two for reagent bottles. Aki and Nikolai worked back to back at the U-shaped end of a pair of benches. At the base of the U were their desks, side by side, perpetually piled with notebooks, blue ring binders, hybridization films, and reference books. There was a jumble at eye level, too, where generations of previous postdocs had taped up cartoons, film schedules, cryptic scrawlings, left there to fade long after the people were gone.

Above the desk, a large window ran the length of the room,

always grimy, more screening out than revealing a stunning view of the Hudson River and Manhattan to the south. I never saw Aki look in that direction, except to hold his hybridization films up to the light.

He had the same series of steps to do, over and over, hundreds of times: Blot gels, label probes, hybridize, develop films, analyze. And then there was the housekeeping. Aki began one day shortly before Christmas by using a detergent to strip labeled probe away from several blots. The DNA fragments, wicked up from a gel, were baked hard onto the filter; he could use it again if he washed off the radioactive probes.

Next, he removed his box of probes from the freezer section of the Sears Coldspot refrigerator at one edge of the room. The probes were in Eppendorf tubes, scores of them, stored in three Styrofoam boxes, each about the size of a flute case and molded with one hundred wells just large enough for Eppendorfs. He referred to his notes, to choose some probes for a new analysis. But Bill came in with an air express package, containing some new mouse hybrid cells that needed to be frozen for storage. Aki put off what he was doing, and went away to care for them.

After seeing to the mouse cells, another task was waiting. He set to work filling three square brown Tupperware containers, rectangles the size of cake boxes, with a buffer solution he had prepared earlier. Each box contained an electrophoresis gel, loaded with samples of DNA that had been exposed to an electric field overnight.

From a large cylinder nearly a yard tall, Aki tipped the buffer liquid into the boxes, and then carried the boxes away to stain the gels with a chemical that would make the DNA fragments on the gel fluoresce so that he could see them in ultraviolet light. (He did this later that afternoon, in a darkroom at the other side of the lab floor. Wearing a large Plexiglas protector the shape of a welder's mask, he flipped on an eerie lavender light and took a Polaroid snapshot of his gel, to record where on the thick slab of agarose the non-fluorescent fragments of DNA had relocated in the presence of the electric field.)

After putting the gels to soak in the fluorescing chemical, he returned to his bench. He began to label his Eppendorf tubes with small bands of green tape, marking them with identifying

numbers to match the ID numbers in his notebook. A Howard
Hughes among scientists might have a secretary to do this cler-
ical task; certainly no postdoc would.

A timer buzzed, and he returned the Styrofoam boxes to
the freezer. Then he removed the Tupperware cake boxes from
the platform on which he had left them shaking, gently and
mechanically. He opened them and carefully poured off the buffer containing the stain. Then he replaced it with clean buffer
and stacked the boxes back on the platform, to wash for another
hour.

At one point, almost in passing, Aki apologized. It must be
very boring to watch, he said—the emperor suddenly aware of
his nakedness, the Wizard of Oz caught blustering behind his
curtain.

"The one advantage of seeing patients was that you had
the feeling you were helping people," he remarked. Now he
was rolling a glass rod over a plastic bag again. "In this job, you
don't spend your time with people. You spend it with bottles
and pipettes."

After lunch, Aki reached into the paper avalanche and pulled
up one of his recent films, to ponder it. He had hybridized probe
number 214, one of the "bad" clones that gave no signal, against
twenty-eight of his other probes, numbers 301 to 328. He wanted
to see whether or not they bound to each other.

Aki had come up with yet another explanation for his failure
to get a good radioactive signal on so many of his early hybridizations. Perhaps some of his linking clones contained what is
known as a *repeat*: a particularly common DNA sequence that is
dispersed across the genome. If there was a repeat near many
of the NotI fragments, any probe containing the repeat would
bind with so many bands that the signal on his thirteen-lane
gels would be diffused, spread out all over each lane. The film
would be murky gray.

If a repeat existed, certainly some of the bad clones would
also bind to each other. He had made a new set of gels, to test
clone against clone.

"Strange result," he said, gazing at the film. "Strange
film."

Some of the lanes were empty; evidently not all of the probes contained the suspected repeat. Some lanes did show bands, albeit fairly faint ones, and not all at the same location. The lane representing clone number 312, about which he had comparatively little information, contained a huge, dark blob, near the bottom of the lane—a very strong signal from a very small fragment, or else a mistake.

Cassandra approached, and asked him about the blot.

"I think it is significant to see if they all share the same sequence," she said. "But it's also important to see whether the repeat spans the NotI site. You ought to digest with Sau3A and probe with one of your bad clones." (Sau3A is a frequent cutter, an enzyme that cuts the genome every 250 bases or so and creates relatively small fragments.)

They might have been discussing industrial chemicals, or even automobile parts, not infinitesimal fragments of human identity. "If you digest with Sau3A and NotI, that will tell you if the repeat spans the NotI site," she went on, "which would be interesting."

"Interesting," Aki agreed with a chuckle, "but a terrible problem!"

From her point of view, Aki might have stumbled on a probe that picked up the starting point of genes, quite a useful little piece of DNA. From his vantage point, there might be a contaminant that fouled most of his library.

"Aki's frustrated," Bill Michels said a few days later, as he was replacing clean flasks on a cabinet shelf. "I think somewhere in the back of his mind he thinks he should be done by now, but we have problems—problems with polymorphisms, problems with signal, problems with the cell lines. The problems we don't expect are the purification problems."

He paused for a moment, flask in hand. "If we could just eliminate every problem," he said, "we'd still have six question marks." Aki had extended his visa to March.

At the next lab meeting, Aki went so far as to volunteer a problem for discussion. Near the end of the meeting, he rose from his seat, walked to the front, and clipped another film of probe 214 onto the light board. He explained what it was, and asked Charles and Cassandra what they thought.

Charles twirled his mustache, and considered. "Cute," he said eventually. "Aki, you make a prediction that this bad clone should preferentially hybridize to other bad clones and not to good clones. That makes sense."

"That film also says the repeat is not around the Not site," Cassandra remarked.

"Why?" Charles asked.

"Because then he would have two bands."

Charles was silent again, looking at the film.

"So whadda we do?" he said. Charles Cantor, the consultant, momentarily at a loss.

Aki laughed.

That may have been the moment when the consensus arose in the lab that Aki's problem getting good signals must have been due to repeat sequences. Later, Aki and everyone else who spoke about it worked under the assumption that this was the explanation. But months later, after Aki had left, Cassandra was still not convinced that the repeats even existed.

"That's what makes science interesting!" she said. "What's wrong with this clone? It's something unknown. And actually that's what sort of divides, I think, good playing from bad playing, focusing in on these things." (Cassandra likes to refer to her line of work as "playing.") "They're really titillating. What's wrong with this clone?" She said this fiercely, as if she wanted to shake the clone by the shoulders.

"Some people would just automatically give up. They'd say, Well, it doesn't follow all the rules. It doesn't obey what we expect and I don't want anything to do with it. But it's some of these things that don't work out as predicted that are actually the most interesting science."

It's easier to capture the tedium in this line of work than to convey what is fun for Ray White or play for Cassandra Smith. Most of what goes on seems repetitive and uninteresting. That's probably why you never see an accurate Hollywood movie about what goes on inside a lab: the best of the action takes place inside talking heads.

One of the pioneers of the field raised this issue at a 1989 press conference about the genome project. "We have to worry about the fact that this work is too dull for intelligent people," said James Watson, the head of the NIH part of the initiative. Then he took it back. "Actually, I disagree with that. It's perfect work for intelligent people. We need machines to do the dull part."

Smith and Cantor seem to feel almost a sense of mission about automating the process. "If you think about it," Cassandra pointed out, "it's unbelievable that modern science is done by people holding up Baggies and cutting holes in them." (She was alluding to hybridization.) "This is unbelievable," she said, "that people are talking about a three-billion-dollar project in which this is what people are doing!"

There was no good way to mechanize all the steps of a hybridization, or to create a computer that could analyze something like Aki's thirteen-lane films as well as he could. As they thought about this, Charles and Cassandra began to realize that they ought to find a way out of hybridization completely.

"Just sitting thinking about all this," Cassandra recalled later, "it became very apparent that DNA sequencing is easier than hybridization. So is PCR. You use little test tubes. You're not dragging around six liters of buffer. I'm not very strong and to me it was manual labor to do a hybridization experiment."

While Aki was making hybridization blots, Charles and Cassandra were pondering the details of a brainstorm that, if it worked, might eliminate the need for most of the little tasks Aki had been doing. They waited to reveal it, even privately, until the lab meeting three days before Christmas. Their excitement then was almost palpable.

Once they began thinking about PCR, they realized that it could make a mapping very simple—indeed, automatic. "Applied Biosystems [a major distributor of PCR equipment] is very excited about our strategy," Cassandra announced at the start of the meeting, "because once we make it and show it works, everyone will want to use it."

To use PCR to multiply a DNA fragment, you always need to start with primers—small bits of DNA identical to the last few bases at each end of the fragment. A single primer of known sequence is easy to make, now that there are machines capable of stringing together small chains of DNA subunits. "The machine is downstairs," a staff member explained to me. "You give the DNA to the lady behind the counter, and she makes it for you.''

The primer sequences at the outside edges of any linking clone, if you think about it, are identical to the DNA sequences on either side of a particular NotI site in the genome. Conceptually, the linking clones and the long NotI fragments link up automatically, if you can obtain those sequences, and the NotI fragments.

But there was a problem with the NotI fragments. They couldn't be stored. Most of them were far too large to fit inside either a plasmid or a bacteriophage, the two kinds of genetic ''containers'' that biotechnologists use to insert their DNA samples into bacteria for storage.

Aki had to use mental deductions based on testing mixtures of NotI fragments spread out on gels, because he couldn't keep a permanent collection of NotI fragments to study. They were too large to clone.

For many months, Charles had been making plans to create a new chromosome 21 library, using a new strategy for preserving DNA libraries, something called a *yeast artificial chromosome* (YAC). These artificial chromosomes, first developed in the laboratory of Maynard Olson at Washington University in St. Louis, contained all the portions of yeast DNA necessary to replicate the yeast chromosome. They could also accommodate much larger DNA segments than bacteriophages or ordinary plasmids could contain—up to a million base pairs long. As long as a NotI fragment.

At some point in December, in one of those over-dinner conversations, it all fell together for Charles and Cassandra. Possessing a complete YAC library and a complete NotI linking library, you could make a complete map of the chromosome: segments joined by linkers, like jewels on a necklace. If you could create a complete NotI library in yeast artificial chromosomes, and sequence the ends of each YAC clone and the ends of each linking clone, the two libraries would link up automatically. (See illustration on page 244.)

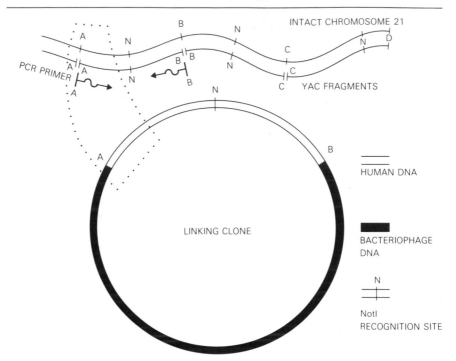

Yeast artificial chromosome (YAC) fragments and linking clones. A PCR primer made starting at site A or B of a YAC fragment will identify the end of a particular NotI linking clone cultivated as part of a bacteriophage. The DNA sequence to the right of point A is identical in every region inside the dotted lines.

There would be almost no need for Aki's kind of mental gymnastics. Furthermore, if they could perfect the polymerase chain reaction and make primers based on all their NotI linking fragments, there would be no further need for hybridizations at all—no radioactivity, no buffers. All they would need to do is to extract the DNA from each NotI clone in a YAC library, and put them one by one into the PCR apparatus with a specific primer for one of the linking clones.

If an end of one of the YAC clones matched the primer made from a NotI linking clone, that NotI fragment would multiply like crazy in the PCR reaction. They'd see lots of DNA in the test tube. If they did not match up, nothing would happen. A hundred such experiments in theory, and you'd have a map of chromosome 21. (See illustration on page 245.)

Simply elegant. A white-glove operation. Of course, adding all those little samples of DNA to all those little PCR tubes would be tedious, too. But that's just the kind of job machines are good at.

"This could eliminate blotting," Charles said jubilantly. "Cassandra has been arguing for two years that we don't want to have to purify anything. With this, you take all the clones on PCRs, and just sequence the ends."

Bill spoke up. "If Aki finds the repeat sequence that he's seeing in his linking clones—"

"If we had that clone," Charles replied, "what we could do by PCR is go into any library, we could make a single copy probe of every linking clone—but there is something else. Let me think." He turned to the blackboard, and drew a few lines. "You could go against an EcoRI partial library and see what lights up, and if there's a repeat compete it away. Anything you pick out with those is something worth sequencing. . . . But you still have to subclone. . . ." He trailed off, and turned back to face the room. One of the technicians was asleep.

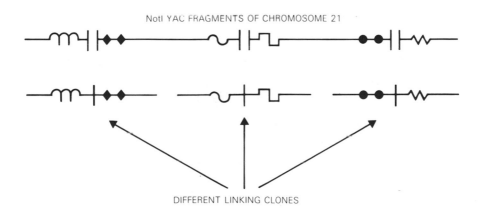

Mapping yeast artificial chromosome (YAC) fragments of chromosome 21 using linking clones.

"Ask a question," he said to Cassandra. "You look bored."
She just sat there. He looked at her, arms crossed.

"Of course, we have to see if it all works," he added, at a
lower pitch.

"So we're basing everything on PCR sequencing, which we
haven't done yet," said Bill.

"Right," Charles replied. "But there are people who promised to teach us: ABI [Applied Biosystems]. They want to sell
sequencing machines, and if this works, they will . . . The ideal
situation is to get the PCR to work so we can sequence the linking library."

They also needed a source of pure chromosome 21, to make
the YAC library, and that would be equally difficult. "When we
ask for a microgram of chromosome 21, they tend to faint," he
admitted. "A tenth of a microgram takes two days."

"What I want to do is prove the principle," he added. "If

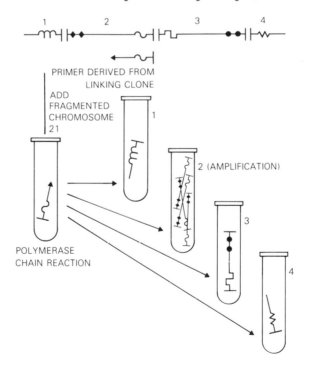

PCR hybridization. Because the primer for the PCR reaction is derived from a
linking clone that matches YAC fragment number 2, only that fragment will be
amplified (multiplied during a chain reaction). The PCR tubes themselves will point
out which fragment the linking clone identifies. PCR does the mapping.

we prove the principle and just link them up, it's history. All we have to do is to be first."

Privately, Charles later referred to this statement as a deliberate strategy to energize the crew. As part of a pep talk. But Jan-Fang Cheng, who had used a YAC to clone his fragment of a telomere, was skeptical after the meeting.

"There's a lot of risk in YAC cloning," he said. "The DNA is too large. On a bad day, you get no clones. There's a lot of trouble in doing this. We knew this from the beginning.

"After I learned YAC cloning," he went on, "I gave it up. I'm still doing the YAC vector, but only for telomeres." Fang had only to make one, or a few. Not dozens or hundreds.

(Four months later, I reread this conversation to Fang, from my notebook. "But that's not true anymore!" he said. "There have been a lot of advances. The vector's getting a lot better and a lot easier to use!" But was it true then, I asked, around Christmastime? He thought for a moment, then smiled, and nodded.)

"The technique has been slow to come up," Ray White had said of YAC cloning at a seminar the previous summer. "At the same time, it's a difficult technique, and a characteristic of difficult techniques is that they get easier."

Soon after Aki returned to Japan, Cassandra revealed the strategy at a scientific meeting. Shortly after that, or perhaps even beforehand, other labs began trying similar approaches, using PCR primers, essentially, as probes. A scientist very prominent in the field assured me this was not scientific plagiarism: to anyone who knew about gene mapping, and to anyone who knew about PCR (and who didn't?), it was almost obvious to merge the two ideas. Obvious to senior scientists, anyway, he went on. The postdocs, perhaps, have less time to brainstorm.

Outside the door that led into Aki's lab, a bulletin board and a blackboard hung on the wall. Occasionally the blackboard was used for general messages. More often, collaborative works of humorous art or satirical prose appeared there, unsigned.

In January, the blackboard contained a list of ten not-so-facetious reasons for being banished from the lab. It vanished, erased by the person who started the list, just before Charles returned from a trip. It was replaced by a whimsical cartoon titled "The Mapping Project." There was a sign that read PICNIC SITE, a large flag labeled NOT SITE, a fence representing repeat sequences (not identified; you were supposed to intuit that), and a lake with a G-C island (the kind of region where NotI sites arise) in the middle.

When that cartoon disappeared, Cassandra pleaded with the postdocs to restore it. They did not. In its place arose "The Other Mapping Project," a phantasmagoric zoo of bizarre animals. It looked like a Hieronymous Bosch. One of the creatures, near the bottom, was a repugnant monster with a huge hole in its head and eight long tentacles. Below it were the words: "Postdoc. Empty head. Many fingers."

Two weeks after Christmas, Aki's experiment had changed course. He had abandoned his search for the purported repeat sequences, and given up trying to account for his bad clones.

"Things have changed," he told me. "I hybridized all my clones with one linking clone, which had a lot of repetitive sequence—a bad clone. Many of them hybridized, but the intensity was not so strong. . . . So I'm not sure how [much] repetitive sequence they have. For now, that experiment is suspended."

It was a management decision, Cassandra said later, because Aki's time in the lab was limited. "I mean, we wouldn't throw it aside. We haven't thrown it aside. But given that there was a time constraint, we felt, let's concentrate on the probes we have data for. That's a decision of playing, and keeping your goal in mind." He was proceeding toward the original goal: to find as many true linking clones as he could.

By midwinter, Aki was not the only person in the lab who moved about quickly and spent little time in idle chat. Many of the postdocs were at least interviewing for positions in other laboratories. Some could say for sure where they would be going when the lab closed in July.

No one besides Cassandra had committed to joining the Berkeley lab. Everyone had a different reason: I don't want to leave New York. I don't like California. I don't want to be part of big business, I want to be part of a university. And also, in some cases, I don't like the management.

"We lose most of the laboratory," Cassandra said in April. "That's our big problem. We've already lost most of the laboratory. We're in a holding action now."

By late February, Aki had begun to piece together the start of a rough map of parts of chromosome 21, testing his best linking clones against a number of cell lines using PFG electrophoresis. "This week, I just used the true linking clones as a probe against everything," he told me, against all the major cell lines at hand. "We are getting some data."

Three hundred clones; six months later, after bad signal, false clones, contamination problems, fragments not on chromosome 21, and all the other exigencies, he could claim fewer than twenty. He had begun to assign his true linking clones to various general regions of the chromosome, by testing them against several kinds of hybrid mouse cells. Each hybrid contained a different large fragment of human chromosome 21 localized by microscopic studies in other labs, so he could begin to pin his clones to a large region of the chromosome, at least.

If these experiments showed that two linking clones were in the same region of the chromosome, and if his blots showed that they both bound to the same NotI fragment, he told me, it would be fair to deduce that the two linking clones represented the opposite ends of the same large fragment. He could put two of his clones in physical relation to each other. That's mapping.

Or, to put it as he did: "If this is proven, we may be able to do some physical mapping. This is not complete restriction mapping—it's very rough. But maybe we will have some anchor points for complete mapping."

By then Aki had identified seven new linking clones, and was teaching his process to a new technician who could carry it on after he left. But he was also preparing blots to take back to Japan, so that he could continue to test some clones himself. If he found any new linking clones, he would send them back.

"At least we have identified ten true linking clones, and we are getting seven more," he said. "Finally, of seventy or eighty there will be at most twenty-five or something."

"That's all I can do using my library," he added, simply. "That's all the capacity of my clones."

■　　■　　■　　■

On March 2, late in the afternoon, almost everyone in the lab crowded between two benches to wish Aki farewell. As office parties go, it was on the simple side: a small glass of modest champagne, then beer, crackers, and cheese. Charles was absent as usual; Cassandra was there, feeling voluble.

The conversation quickly turned to competition. Cassandra spoke about a time a collaborator had shown a slide of one of her blots at a scientific conference, without giving her credit, and how she had called him on it later, privately. At first he denied that the data were hers, she said—until she told him she could easily send him a photo of the original, at which point he relented.

"I moved him from my good list to my bad list," she said. "I really have lists." She eventually moved him back, she added, but only after he called her and—she searched for the word. *Groveled,* offered a technician. That's right, she said. Groveled.

Everyone chooses whom to collaborate with and whom to avoid, she told me afterward; anyone would "be regulated by" such an incident. "There was nothing he could do to make up for the good that would have been done if he had given me the credit in public to begin with," she explained. "And he certainly got credit."

She had learned to be cagey in advance. At an important upcoming meeting, she said, she would be using the internal lab names for their probes, not the universal names. And instead of printing the fragment sizes as numbers, she added, she was going to show them schematically, drawn to scale next to the chromosome. For most people it wouldn't matter; for the competition, it would be inscrutable.

She said it with glee at the time, but later Cassandra argued to me that it would be considerable work to relabel all the probes, and that "the map was not finished and was simply a work sheet. People do not understand maps are not done until they are finished. . . . If one shows incomplete maps which have the appearance of being done . . . everyone tends to think the map is done."

There's nothing unusual about all this, except perhaps her unguarded candor at the party, in the presence of an outsider. Most scientists want their labs to be first. Given what's at stake in the genome project, it's not surprising that, as James Watson once lamented, "competitive races dominate what should be a noble point in human history."

Cassandra had special reason to be worried about competition that afternoon. At that moment, Norman Doggett was visiting his future employer and present rival. He was under strict orders, Cassandra said: Anything they want to know about the research in New York, they call me. Norm was to say almost nothing at all of what he had been doing.

She teased Aki in the same vein. Going to Japan now, he might also become a competitor. After all, two of the five groups trying to map chromosome 21 were in Japan.

A few months later, James Watson would allude to the special competition surrounding chromosome 21. "Since [it's] so small," he said, "a lot of groups want to work on it. We need to face the problem of how to kick somebody off. Maybe this should be the example of where the jungle works—so everybody competes on chromosome 21, and we're civilized on the others."

Aki was acutely aware of this competition, and it had begun to worry him. He told me that his recent experiments showed that some of his NotI fragments were closer to 3 megabases long than to 1 megabase. So it might take nearer twenty linking clones than fifty to map the entire chromosome. He might already possess the complete collection needed to link up all the NotI fragments. At first this cheered him, until he began to wonder who else had as many linked clones as he had—and for how long they had had them.

The heartening news was that Cassandra and Charles had asked him to write up the linking clone results for publication. What troubled him then was the thought that someone else might beat him to it.

From the start, the understanding was that the linking clones would belong to the Cantor lab. But Aki asked Charles and Cassandra not to work on the hypothetical repeat sequence, and they agreed. "Many times, when postdocs leave the lab, they will take some part of the project with them," Cassandra explained afterward.

Aki left the library behind, as agreed, and also the data base to explain his knowledge about each clone—something so dense with information that no one claimed to understand it very well. Until the last moment, he was racing.

"He was working like a dog before he left," Cassandra told me. "At some point, I just had to stop him and say, Forget it. You know he's a good scientist, and he had certain goals he wanted to accomplish. It was admirable that he was willing to work day and night, but at some point it was more important that we understand what he was leaving behind than that he try to finish up the last couple of experiments. I had to get him to stop this panic."

At the time he left, she said, Aki showed her a long experimental plan outlining what he still intended to do with the clones and blots he was taking along. "It was so unrealistic," she said. "I was kidding him. I asked if he was gonna do it on the plane."

The longer arm of chromosome 21 contains, among other things of peripheral interest to Akihiko Saito and Cassandra Smith, a gene linked to Alzheimer's disease. Someday, not long from now, there will be a fairly competitive physical map of it, a linear sequence of NotI sites and BamHI sites and EcoRI sites and more, with landmarks much more precise than the ones that appear on the existing map, the one made with genetic linkages.

For what is ultimately known about chromosome 21, everyone will owe some small thanks to Akihiko Saito, for all his labors, mental and manual. But it's not a good bet these labors will win him any great renown. That kind of recognition goes to the likes of Charles Cantor and Cassandra Smith. Aki was just, after all, a member of the crew.

A brief postscript is in order. For the laboratory in question, the tug between the academic life and the needs of the genome program was temporary. Long before Charles Cantor and Cassandra Smith moved to California, people had begun to refer to their operation as a factory. Charles Cantor himself sometimes described the future Berkeley lab in those terms. As perhaps it should, the physical mapping project was evolving to resemble a small business more than a university department.

There's an interesting message in the history of one peripheral character, a biochemist named Stuart Fischer, whom Charles recruited to the Berkeley lab during a Columbia departmental retreat in the autumn of 1988. Fischer was there representing a small biotechnology firm, Lifecodes, which had provided money for the retreat.

Stu Fischer had known Charles Cantor since 1970, when Stu had been one of the students in his undergraduate biophysics class. "It was the kind of stuff most people thought of as you wanted to avoid," Stu recalled later. "It was perceived as being difficult—mathematical, chemical. But you knew that this was quality learning. . . . He conveyed an excitement and a purpose about the subject, and did not seem to be bored by it, did not seem to apologize for it. He was dynamic. His excitement made me want to learn it."

Fischer did learn it, and went on to do the kind of work Charles Cantor had been teaching about. By the time of the retreat, Fischer had gained a Ph.D., and afterward had chosen an alternative to publish-or-perish. He joined a biotechnology company. At Lifecodes, he had been promised, the genetic research effort would expand as soon as the company got a foothold in the business, using genetic markers for paternity testing and forensics.

But there are pitfalls in this path, too. The paternity business spun money, all right; Lifecodes was bought out by a multinational corporation, which was phasing out the research arm and wanted to scale up the paternity and forensics operations.

Charles Cantor offered Fischer an escape. He invited him to join the Berkeley lab. Fischer accepted.

A few months later, when Lifecodes was embroiled in a major controversy over the quality control of its DNA testing for a murder case in New York, Stu Fischer was working as a mid-level manager in the Cantor/Smith lab, safely out of business but also safely out of the ivory tower.

"Academics love technology," he told me, "but most scientists do not like to be known primarily for developing techniques. They like to be known as biologists, interested in the process of life. People interested in developing technologies sometimes don't—" he paused to search for the best words— "don't fit the classical mold for the academic scientist."

One got the feeling Stu Fischer would not be at all insulted to be called a technologist rather than a scientist. ''Most people who develop techniques do so because they had a problem they were interested in exploring,'' he said. But then they are remembered for the technique, not for solving the problem. (Charles Cantor, he failed to add, is a case in point.)

After all, he went on, technical developments are the things that expand our ability to explore.

10

FRANCES

In a few years, probably, we shall know the identity of most of the human genes. What difference will that make? Given the opportunity to know about your genes, would you really want to know?

Some of the most poignant and crystalline answers come from the people at risk of Huntington's disease. New genetic tests provide them with the opportunity—and the curse—of being able to learn their fate years before they could expect to confront it.

Choose another name for yourself, I suggested. She thought for a long moment, and then said, "Frances."

Frances lives in a well-trimmed suburban neighborhood of large homes on broad lawns. Her directions lead to a substantial colonial house on a corner lot, set well back from the street.

Frances came to the door smiling, a slender woman in a tailored dress, with refined features, dark hair, a bone-china complexion. As she made fresh coffee and heated croissants for us, a little girl tumbled on the back lawn and a small boy, a toddler, embraced her lower leg with one arm and stretched upward with the other.

At length, Frances called the au pair to watch the children and then, carrying the tray herself, led me through double doors into the living room. She set the tray down on a coffee table between two sofas. Then the phone rang, and she left the room.

Waiting, I read the spines of the books in a low bookcase: Standard medical references (*Harrison's Principles of Internal Medicine*, *Stedman's Medical Dictionary*, a surgery text), nearby several shelves of legal texts, and above those, a substantial collection of theology and philosophy books. Who lived here? I knew nothing about her but her real name and that she had been tested for her fate with regard to Huntington's disease.

Frances returned, and closed the doors behind herself. She poured coffee into china cups, and we began to talk, facing each other over the coffee cups. Except for changes needed to disguise her, all that follows is true.

Frances first became aware that she was at risk of Huntington's disease when her father was diagnosed in November 1984. The first obvious sign of his disease—though no one in the family recognized it as such at the time—had been a psychotic episode that sent her father to the hospital for three months in 1974. Shortly after that, he began to develop strange movements of his arms and legs. Psychiatrists said these were due to tardive dyskinesia, a side effect of his psychiatric medication.

His first hospitalization ended in late 1974, and he went home. He stayed there for ten years, under his wife's care, until he had another psychotic episode. The family had him committed again, to a suburban psychiatric hospital. Soon afterward, he began to have seizures. He was rushed to a medical center in Baltimore, where his condition was finally diagnosed correctly.

In retrospect, it was clear that the disease had begun to emerge many years earlier. Frances's best memories of her father are distant ones, of dinner conversations when she was a child. The oldest of three daughters, she did very well in school. He was very proud of her. He would ask about school, and then listen attentively to her chatter about schoolwork and classmates.

When Frances was twelve or thirteen, things began to go sour. Her father became difficult. On family outings to a restaurant, for instance, he would order for everyone without asking what they wanted. "He couldn't face the management problem," she explained.

He also had gross memory lapses. Frances's mother worked very hard to arrange a high school graduation party for her. All of a sudden, her father was furious about it. He said no one had ever discussed it with him, although Frances remembers her mother talking about it a great deal. He had totally forgotten, or at least claimed to. Then he prohibited the party, said he would not have all those teenagers in the house. His wife had to call everyone and cancel.

Frances's father was a minor executive with a large privately owned company near Washington, D.C. In the late 1960s, the family learned much later, people at his firm had begun watching him for strange behavior, secretly evaluating his mental status. The family never learned the details, but in the summer of 1974, after twenty-five years at the firm, he was fired.

On that day, Frances's father came home and "started ranting around," she said. "He would just sit on the sofa like this, and talk, and talk nonstop, about what he was going to do, how he would sue the company, sue this person, sue that person. He would pace. He'd get up and sit down. He was just nuts. You could tell."

At the time, Frances was a summer associate at a law firm in Washington. She can no longer remember why, but she was at home that day. Her father said he wanted her to get her firm to sue his employer. She was twenty-three years old, and terrified that he would call the firm. "Well, this was my future!" she recalled. "I liked the firm. I wanted to come back. I was frightened. So I called one of the partners, and told him to accept no calls from my father, that he was very sick, that something dreadful had happened."

Her mother went to a neighbor who was a social worker, and the neighbor agreed to try to help get him into a hospital. After several hours, a truck came from the hospital, bearing people in white coats. They walked into the house and put her father into a straitjacket. He objected, but they just said, "You're going with us now." He followed.

When he returned from the hospital after three months, Frances's father was subdued—"very sick and broken apart," as she put it. Most of the time he just sat and twitched. He'd put his foot up and it would twitch. He couldn't stop it, and it was nerve-racking.

Meanwhile, he was full of demands: Get my slippers. Get me an apple. Just like a three-year-old child. Frances saw all this, because she lived at home then, during her first year after law school. She was working at a firm and waiting for her fiancé to finish medical school. They were married in 1976.

Frances's mother endured for a decade, caring for her debilitated husband and working part-time. Finally, ever less able to cope and ever more depressed, she began seeing a psychiatrist. She knew this would infuriate her husband, so she kept it from him. The psychiatrist kept pressing, saying, "You have to tell him."

"He'll freak if I do that," she insisted. Finally, after a few weeks, she did tell him. And he freaked. He gave her no peace about it. He brought the subject up all the time, and he would telephone people and tell vicious lies about the psychiatrist.

It didn't take long after that for the family to decide to have him committed again. That was June 1984. Late that year his condition was diagnosed correctly, but he stayed at the mental hospital until a bed opened up for him in a nursing home, in February 1985. It will be his last home.

Huntington's chorea, as the disease is sometimes known, is an incurable, inexorably progressive and fatal disease caused by brain degeneration, most notably the premature death of nerves in the basal ganglia, the gray matter that is involved, roughly speaking, with movements, mood, and thought. The disease was first described in 1872 by Dr. George Huntington, who found a large affected family living on Long Island in New York. It is inherited in an autosomal dominant pattern: the incidence (roughly 1 in 10,000 people altogether) is equal in both sexes, and anyone who has an affected parent has 50–50 odds of inheriting the disease.

Huntington's disease usually manifests itself for the first time at some point between the ages of thirty-three and fifty. It may begin with unusual, involuntary movements or with prominent psychiatric abnormalities: loss of memory, judgment, and other mental faculties. Severe depression often predates or accompanies the development of other symptoms. Irritability and explosiveness are also common.

Although there is a fair amount of variation in the age of onset and the severity of the symptoms, Huntington's disease inevitably progresses toward dementia. At the same time, it also leads gradually to loss of control of facial muscles, unpatterned movements of the limbs, faulty speech, irregular breathing, and, eventually, in most cases, to near immobility. Despite many biochemical studies of peripheral tissues of living patients with Huntington's disease and of brain tissue from its deceased victims, there was no clue to the primary cause of the disease at the time Frances's father was diagnosed—or, in fact, at the time these words were written.

Just before Frances's wedding, a distinguished neurologist wrote in a differential diagnosis that, among other possibilities, her father might have Huntington's disease. Jeff, her fiancé, heard about it. At the time, he was a fourth-year medical student. He and Frances had been together for more than six years.

Jeff was petrified. If it was true, what could they do? He had always said it was one of the meanest tricks that life could play, to put something like Huntington's disease in someone's future, without revealing it until after the children are born. And now it might be in his wife's future.

Frances's mother told Jeff that it wasn't at all that definite, that Huntington's was just one of a number of possibilities. Perhaps he wanted to be persuaded. For some reason, he took her word for it: her husband had tardive dyskinesia.

They went ahead with the wedding. In 1982, Jeff went into private practice, and Frances continued to work. Then she got pregnant. Within a few days after she learned she was pregnant, and before she had told anyone but Jeff, Frances had a phone call from her mother. The doctors were now saying the diagnosis might really be Huntington's, her mother said.

"Well, no, it isn't," Frances replied.

Within a day or two, the worry preying on her, Frances called the doctors herself. "How have you decided this?" she asked. "We were always told it was tardive dyskinesia."

"The movements don't look like tardive dyskinesia," the doctor replied. "The entire pattern of movements is different. It's not that." With a porcelain clink, she set down her coffee cup.

Of course, she told Jeff about it. "This could be the last time we are without this problem!" she said. "If this is it, can you *imagine*"—she chuckled lamely—"what this is going to mean for the rest of our lives?"

The family got a second opinion, from another medical center, in November 1984. November 5, a Monday. The verdict was the same: Huntington's was a distinct possibility.

The dates are like scars in her memory. Friday, November 9, was the first time she actually believed it was Huntington's. That was when, through a little subterfuge, she discovered that it ran in the family.

At about two o'clock that afternoon, she placed a call to Indiana. Jeff had the day off. They were upstairs in the bedroom, talking about Huntington's disease.

The second opinion had jolted her into learning whatever she could about the disease. One of the many things she learned was that a confidential roster in Indiana kept track of families with the disease.

"You know, there's this roster," Frances said. "If they had my maiden name in their files that would tell us a lot, wouldn't it?" It's a very unusual, multisyllabic surname.

"It certainly would," he replied. They looked at each other, and she picked up the phone. ("You don't wonder about something like that," she explained.)

A young man answered, and she described the situation. There was silence as he scrolled down the computer screen. After a few moments, he spoke. "Um, yes, we have that name here. It's just a little tiny family down South. We have very little information about them." Then there was another silence.

"Are you sure that's the name?" Frances said, and spelled it again.

"Oh, yes," he replied. "That's what it is."

Frances knew of only one family on her father's side that lived in the South. She called the next morning.

A man answered. Frances explained who she was, and told him about the roster. She said, "We've had a diagnosis of Huntington's disease. Do you know anybody in the family who had this?"

"My father died at the age of sixty-five," replied the stranger on the other end of the line. "There was a diagnosis of Huntington's disease somewhere in there, but the doctor decided that it wasn't that. They thought it was, but they established it definitely wasn't."

"Well, it looks like my father has something like that," she said. She can't remember how the conversation ended. It was brief.

Within less than two hours, she got a phone call from his mother. "She was frantic," Frances told me. "She said, 'Who are you? How do you know you're related to us?' I said, 'I am so-and-so's granddaughter. He had a brother, Meyer, and a son, Meyer, so the connection's direct and undeniable.' "

The woman in the Southern branch of the family had six children, all of them adults. For years they had lived with a diagnosis of Huntington's disease, and then the doctor had told them not to worry. They put it out of their minds. Then out of the blue, over the phone, a total stranger put it back.

Frances remembers sitting at the dining table with Jeff that evening, crying over her food. Jeff was at the edge of tears himself.

"This is awful," he said. "This sucks! This is never going to go away! We'll be with this for the rest of our life. There is nothing we can do."

But he was wrong. Within days, Frances learned that she could be tested for the Huntington's gene. The testing hadn't been set up yet, but it was going to start soon.

Until 1983, there was no way to predict beforehand whether someone whose parent had developed Huntington's disease was destined to the same fate. In that year, James Gusella at Massachusetts General Hospital and a large team of collaborators reported finding a genetic marker for Huntington's disease on the short arm of chromosome 4. It was the same finding that spurred Roger Kurlan to seek out a family with Tourette's syndrome and energized the two teams searching for a genetic marker for bipolar disorder among the Amish.

It was the first human disease to be mapped using the new methods of DNA analysis. The study began, as you will recall, with the discovery of a large kindred affected with the disease,

in a fishing village at the edge of Lake Maracaibo in Venezuela. A Venezuelan physician had traced the kindred, numbering
over 7,000 people, back to a single immigrant who had the disease. This implied that, at least in this large, isolated kindred,
everyone with the disease had the same genetic defect.

In 1981, Nancy Wexler and Thomas Chase, then of the National Institute of Neurological and Communicative Disorders
and Stroke in Bethesda, Maryland, traveled to Lake Maracaibo
to collect blood. Initially they were hoping to find a person who
had inherited a double dose of the Huntington's gene, in order
to search for a particularly obvious biochemical defect. Soon afterward, the idea of human linkage mapping using RFLPs came
to light. They returned to Venezuela every year, to repeat neurological exams on the members of the kindred and draw further
blood samples.

In 1982, Gusella began testing DNA samples from the families with a randomly chosen collection of human DNA restriction fragments. He was phenomenally lucky: one of the first
RFLPs known to exist in humans showed a linkage to Huntington's disease. Gusella called it G8. Genetic analysis suggested
that the G8 marker was some 3 to 5 million DNA subunits away
from the gene.

Subsequent studies revealed genetic markers closer to the
putative Huntington's disease gene and also established that
those markers were equally valid in large American families affected with the disease. But though the G8 marker tracked with
the gene in most large kindreds with Huntington's disease, the
exact pattern varied from family to family. In some cases the
disease coincided with one variant of the polymorphic DNA
fragment; in other families it would be inherited, reliably, with
another variant of the G8 marker. (It's as if in some families only
blue-eyed people developed the disease, and in others it coincided only with brown eyes.)

Until the precise gene came to light, therefore, it would be
necessary to test many family members with and without the
disease to determine whether a certain individual would develop Huntington's. Without the gene itself, the estimate of risk
could never be 100 percent or zero. There would always be some
possibility that by chance, during reproduction, a recombination
had switched a region of two chromosomes between the G8

marker and the disease gene, linking the disease in that individual with an alternative form of the marker.

The closer the marker was to the gene, the less likely recombination would become. Newer markers discovered within five years of the first linkage allowed genetic counselors to predict the risk of Huntington's disease to an accuracy of greater than 99 percent.

However, progress toward finding the gene itself was disheartening and slow. The gene lay somewhere between the G8 marker and the end of the short arm of the chromosome, its telomere. The only practical way to pin a gene down using molecular methods was to surround it, to find flanking markers on both sides. At the time Frances's family learned the truth about her father, there were no good markers at the distant tips of chromosomes.

By the time Frances told me her story, telomeres had been isolated in several labs and the gene seemed almost at hand. But when I spoke to her, Frances seemed not to fret about the slim odds that, in her case, the result was misleading.

"Why would you want to be tested?" her sister Madeleine asked Frances.

The reasons seemed so compelling. "You may not have it!" Frances replied. "It may be totally unnecessary to worry." There seemed no alternative. She ignored the impact that the wrong knowledge might have on the rest of her life.

If she died tomorrow, Frances continued, or even back then, she would have had a wonderful life. She was looking forward to the joys of motherhood. She had an interesting career, a stable marriage, and all the money she could imagine wanting. If the news was bad, she would be cared for.

Her two sisters had none of that. Madeleine worked as a secretary, had a low salary and no job prospects, lived alone as she approached middle age. Jennifer, the youngest sister, was a wife and a mother now. Her husband, who was a high school teacher, wouldn't talk about the problem. Frances had decided they were "emotionally delicate" on the subject. They had a one-year-old son, and Jennifer was pregnant. She has told Frances she will not have the test until there's a cure, and that's almost the last she has said on the subject.

"They will insist that it does not affect them, that they live day to day and everything is fine and they put it out of their minds," she went on. "I don't believe it for a moment." She repeats the sentence slowly. "I don't believe it for a moment. You would be unconscious if you didn't put energy into denying it, keeping it in the back of your mind."

While the genetic test for Huntington's disease was being developed, surveys of people at risk suggested that three out of four would take the opportunity to determine their fate. But when push came to shove—when the Huntington's disease clinic at Johns Hopkins Hospital in Baltimore began offering the test in late 1986—a far smaller proportion actually took advantage of it.

By March 1987, only about 13 percent of at-risk individuals notified about testing actually went through with it. Perhaps this is not surprising. They had one of the most poignant, potantially tragic opportunities a person could face.

Long before that, researchers in the Huntington's disease clinic at Hopkins had begun to ponder the psychological and ethical implications of predictive testing. By late 1986, they had funding from the National Institutes of Health to carry out an explicit study of these matters.

Anyone who came to the clinic for predictive testing had to agree to undergo a large battery of psychological tests. Those taking the psychological tests included some at-risk volunteers who had decided beforehand that they did not want genetic analysis but were willing to act as controls in the study. These tests were aimed at trying to determine the qualities of people who sought predictive testing, the effects of an adverse result, and also, in time, the characteristics of people who managed best after learning bad news about their future.

The Baltimore clinic was the first in the country to offer predictive testing for Huntington's disease, but by the time Frances went for her tests three others were in existence, in Boston, New York, and Vancouver. (By 1988, when Integrated Genetics Inc. began providing the DNA analysis for Huntington's disease commercially, there were four additional centers in the United States: in Florida, Minnesota, Michigan, and a second in New York State.)

The original four centers for predictive testing of Huntington's disease shared several important practices. Before testing, they tried to ensure that any individual at risk had a specific support person, called a companion or confidant, who would accompany the person through the testing, and help him or her deal with the consequences. No one who showed symptoms of serious depression or other severe mental disorders before testing would be allowed to go through with it until the problem could be resolved. Children and adolescents, regarded as being incapable of truly informed consent, would not be tested.

"We are the first to admit that when we started the project, we thought we were smart cookies," says psychologist Jason Brandt, director of the studies at Johns Hopkins. "We figured we had thought of every possible ethical issue."

Straightaway, they had to face some unexpected and knotty questions. Someone in the program needed to get a blood sample from a particular relative who lived far from Baltimore. The relative was intensely eager to learn his own fate. But the client population was limited to people living within a reasonable distance of Baltimore, in order to accommodate the exhaustive psychological counseling that was part of the Hopkins program. This man was not part of the program; he was tested purely in order to resolve his relative's future. He was not undergoing the counseling. His test result came out normal; he was not at risk. Should they put the man at ease?

I asked Brandt what the psychologists decided. He responded by holding an imaginary key to his lips, and turning it. "That hurts," he said. "I sometimes have this fantasy of making an anonymous phone call in the middle of the night, saying, 'Listen, you're not at risk.' " But if he began doing that, he went on, everyone he did not call would wonder why.

In another family tested at the Hopkins clinic, a man whose wife had Huntington's disease was asked to give a blood sample so that one of his daughters could have predictive testing. He agreed to do so, but only in her case. None of her siblings, he felt, would be able to deal with the consequences.

At first the father did not understand that genetic blood tests are not like blood tests for an infection: The result from a single test is valid forever, and cannot be forgotten. But the father wanted the information available only to his oldest daughter. Finally, the Hopkins clinic signed an agreement to draw his blood on the condition that the results be limited to testing for his eldest child, unless he gave written permission otherwise.

In a third case, a man wanted to know his risk, but his identical twin did not. However, as soon as one knew, the other would know by definition. Who had the right to decide? "Our working rule," Brandt said, "is to postpone testing someone if doing so forces unwanted genetic information on someone else." Still, he worries about the potential for testing under duress.

The predictive testing program was due to open in September 1986. In the spring of 1986, unexpectedly, Frances got pregnant again. "We felt very stupid," she admitted. "But we were delighted, because we wanted another child."

During the summer, Frances immersed herself in the matter of Huntington's. She went to a convention in the Midwest, and met many people involved with the disease. She visited her distant relatives in the South, and also a distant uncle in Nevada who was now very ill with Huntington's disease.

It was overwhelming. "He had shriveled up," she recalled. "He was a man of about five foot ten, and he'd shriveled down to five foot six and one hundred twenty-five pounds. He couldn't speak; he could look, you knew he understood what you were saying, but he didn't respond. It was really a strain for him to get a word out."

Altogether, it made her very edgy about the test. "You've been impossible—snappish and irritable," Jeff told her afterward. "I'm quite sure it's the issue of starting the testing. Why don't you think about it? Get some counseling.

"You're pregnant now. This is the most immediate thing. We're gonna have the baby. Put it off. Have the baby. If you feel the same way afterward, you can do it then."

Frances did put it off, and she had a few sessions with a psychiatrist. "I kind of put it on the back burner," she said. "As far as it could be."

How far is that? I asked her.

"It's always there," she said. "It was just less bad that time than at other times. If I would put my hand up and bang something, or if I'd forget something, it was there. Every time I'd forget something or trip, I'd say, oh, this—this could be it. I got used to it."

Christopher was born in January 1987. By the time he was six months old, she felt ready to start testing. She wanted more children, and she wanted the truth. This very fact, that she could face the truth, made her feel good. It was one of the few things in recent years that made her feel distinguished.

By the time she was a mother of two, she had begun to regard her career as a steady slide downhill. She had been very good in school: at the head of her high school class, a Vassar undergraduate, and a Harvard law school alumna. Then she had practiced law for eight years while earning a master's degree in tax law at night. "I've got three degrees after my name," she told me. "I invested a lot in myself."

But then she couldn't seem to stay with any job. After three years at her first job, in the trusts and estates department of a sizable firm, she began to feel too confined, so she left to join another firm. After a year there she was dismissed; they felt her expertise didn't match her salary. She went to a third firm, and again was fired after a year.

For about ten months, she tried to find a new direction. She read books and took career-planning courses. Nothing came of it. Finally, she joined a small-town law firm near her home and, to her surprise, enjoyed the work. Then she got pregnant, for the first time. After her daughter's birth, with an affluent husband and the shadow of Huntington's disease over the family, she had not returned to her job.

By the time she had her second baby, she was feeling increasingly mediocre and incompetent. She meant to be a happy housewife and mother, but her new role failed to provide her with a strong sense of accomplishment. "Pursuing the Huntington's test, that did," she said. "It did.

"For one thing, it's the supreme challenge of your life. There's this horrible thing moving inevitably toward you, this monster, this precipice, this—something that is so frightening. I saw what happened to Dad. To look back at all those years of emotional disturbance and not to know that this is it, and that really he couldn't help it, and he got worse and worse, more rigid, impossible, forgetful. To see this thing happening and not understand until twenty years later that he should have been a different person, that's horrifying.

"Then to think that this is incurable, and this could happen to you. I kept saying, Are you going to face it? Are you gonna turn and run away?" She repeats the words slowly. "Are you going to turn and *run* from that? You should know the answer."

Asked to guess, Frances would have said the answer would be bad news. To know that answer would be to master her life. She would be alert for the first signs. She would not resist treatment as her father had; she would do anything to preserve her relationship with her family. She would look the monster in the eye. She would not run.

Frances's testing began in September 1987. It consisted of five sessions. Jeff was her confidant; he came along every time.

The first session was relatively brief. A doctor and a social worker met Frances and her husband, and gave them about two dozen pencil-and-paper questionnaires. Why did they want to be tested? What was their family background? Were they interested in the library, in sports or in music or art, did they have a lot of rules at home? It was easy.

The second visit was simply a long interview with a psychologist. She asked Frances lots of general questions about her childhood and whether she had a family history of mental illness. Then she began on Frances herself, whether she had ever had emotional problems or had reason to consult a psychiatrist, and why.

"I have on a couple of occasions consulted a psychiatrist," Frances told me. "In each case, I was having a lot of problems at work, and I was depressed. It had been years since I had thought of that whole thing, and it brought a lot of failures back." She sighed.

The third session began with a very brief neurological exam. Evidently the neurologist was quickly satisfied that no signs of the disease had begun to develop, at least yet. Then he spent nearly an hour drilling her: Was she certain she wanted to do this? Did she really want to go through with it?

Frances came back with every response she had ever thought of. She told him her feelings about facing the monster.

She told him that she didn't want things to turn out as they had for her father and her. She said she wanted to have more children.

There was another factor too, she told me somewhat hesitantly. "I probably made some reference to the fact that if I had a positive result on the test, my religious faith would sustain me to some degree. But this is a difficult thing to get into. It's private. It's hard to talk about."

She also gave the neurologist a further reason for her decision, which had come to her quite recently: in the same situation, she was certain, this is what her father would do. "He would elect to be tested," she said. "He would consider that it was nonsense not to."

"This was the best of his character," she told me later. "The straightforward, literal kind of thing: Let's know the answer and let's not wreck our lives with wondering, because it's going to get in the way a lot. That was the best of him. There was a certain integrity about him that I was proud of."

During the fourth session, on October 23, they drew her blood sample. Beforehand they asked a few questions about how she felt. Then she stretched out her forearm and gave away a few milliliters of vital fluid. It was like stepping onto a roller coaster.

"I had done it," she said. "Someone was going to know the answer, even if I decided not to at the last minute. The blood sample would be sitting there. The DNA would be drawn out and read."

Sometime in the week after the blood was drawn, a new thought suddenly dawned on her: the result might be negative. She might not have the gene. Wouldn't that be wonderful? What would she do?

Well, I'd cry, she told herself. I'd be so startled that I'd cry. The thought lasted for a moment, and then she dismissed it.

On the whole, she kept the opinion that the result would be positive. She had spent years seeing herself as vulnerable. Also, something told her it would be only fair for her to be the daughter who got the disease.

She had a wonderful life, which she felt she had done nothing special to deserve. This would complete the pattern, she told herself. It would be just. She resolved to be noble about it, but it seemed unbalanced for her to escape. She had been so fortunate.

It's a common fallacy among families at risk to assume that one sibling will be spared if another inherits the gene. In fact, each conception is as unrelated to another as two flips of a coin.

By 1989, early data from the psychological studies at the Hopkins clinic showed that people given bad news from predictive testing were coping surprisingly well, with the support of confidants and the knowledge that they can have free psychological therapy at the clinic at any time. "We like to focus on hope rather than hopelessness," Brandt told me. "We talk about the research toward a gene. We talk about the drug trial we have going on here. We talk about the available therapies." (I wouldn't want to imply that there's a treatment for the disease itself, he added later, only palliative measures, as for the common cold.)

Curiously, although the absolute odds that an at-risk person will develop Huntington's disease are set by nature at 50–50, only 9 of the 38 people whose test results were informative got bad news. That's a proportion of less than one in four. Why is the outcome so skewed?

Perhaps people who are destined to develop Huntington's sense this somehow, before they develop any symptoms doctors can detect, and therefore choose not to be tested. Some other results of the study hint at this explanation. All confidants are asked beforehand to predict whether their friend's result will be negative (unaffected) or positive (affected). Slightly more often than chance would explain, they correctly predict a positive test result although the person at risk shows no symptoms.

Ultimately, the clinic hopes to determine the qualities that help people cope with an adverse result from genetic testing. To date, the subjects of their study are not providing the information.

"We're not able to find people who are doing very poorly," Brandt said, with a chuckle. "We like to think that what we're doing helps." From other studies, it is possible to predict some of the coping factors, he added, such as strong social supports and the absence of psychiatric illness. They have tried to build these factors into the program from the outset. Only one person in the project has been terribly shocked by the result, and disconsolate for about a day, he said. Afterward, she got back on an even keel.

"People tend to put it behind them," he said, "and get on with their lives. They say it gets into their consciousness about once a day. But they don't seem to get very depressed. They choose not to think about it. What's going to happen is going to happen."

On the other hand, no subject who has learned bad news has yet begun to show signs of the disease, and Brandt can't predict what will happen when the prediction becomes reality. There is always the chance, of course, that researchers will find a cure for Huntington's disease, after discovering the gene. No one can name the odds of that, or the time scale.

At some point before that time, the gene will be discovered but not the cure. A company might market a test based on the gene, rather than just on the linkage. There could even be at-home test kits, like the over-the-counter pregnancy tests already on the market.

Someone who hears bad news would no longer be able to grasp at the infinitesimal chance that he is a recombinant, that although he has the marker he still may not have the gene. Told he has the gene, he will know for certain. The test may be offered without the psychological testing, the mandatory confidants, and the family contact that is unavoidable during linkage studies. In such a situation, life might not be a limbo. It might be simply hell.

Looking back, Frances feels that the ground was made ready for her father's diagnosis. During the year prior to his diagnosis, while he was again in the mental hospital, she had felt very restless. Certain questions kept coming to mind, certain anxieties about the disparity between her own fortunes and those of her mother and her sisters.

The anxiety led her to read certain philosophy books. For the first time, in those books, she encountered educated people with backgrounds like her own, with top academic credentials, who had come around to a religious faith and had some persuasive reasons for it.

Since childhood, Frances had always equated religion with ignorance. Her parents were very proud to be nonreligious, she said. "They thought they were enlightened while everyone else was benighted, and I'd been like that too." But before the diagnosis, she had begun privately to read some religious books, and to turn them over in her mind.

Then the Huntington's diagnosis came, and it was overwhelming. It seemed to confront her with an inescapable conclusion, she told me: either there's a God who has some reason for this, or else life is meaningless. Frances could not accept that latter alternative.

She speaks forthrightly about it, when asked, but there is a trace of apology in her words. "I can't justify that on any rational basis, and that's why I am embarrassed, almost, to talk about it," she said to me. "I try, but I don't feel I'm terribly persuasive. I felt that this had to be an event that had meaning for me, that it was a message to me, across time, to which I was called upon to respond."

For several months before the diagnosis, after all, she had been in the process of changing her mind. "It was like there was fertile ground for this lightning to strike," she said, "and then it struck."

She resists the facile conclusion Jeff's mother tries to force on her, that God punished her father for leaving the church and then capitalized on the punishment to reclaim Frances. "I don't like that," she said, "because I don't believe that my father was this doomed, evil person.

"I'm not going to make sense out of it, because I can't. But there's just a sense that there's a meaning to life, and there are reasons why things happen. He's at work there. I wanted to believe that, and I've seen no reason to retreat from that ever since."

Jason Brandt told me he was "amazed by the number of people who say it's their religious faith that gets them through, their faith in God. That's often overlooked." What is the number? I asked. He could not tell me. "We never thought to do a systematic study of that aspect," he said.

A fatal flaw of the genes must force questions on someone that no scientist can answer any better than anyone else. Since the beginning of time, people have been blaming God for misfortunes and seeking Him for comfort. There's no reason, on the face of it, why an infirmity of the genes is any different from any other infirmity with regard to the reason that it strikes one person and not another. The questions are no easier couched in terms of DNA.

Evidently some of the ordinary people, those first to be affected by the new knowledge of our genome, are taking their own counsel from an element of the cosmos that does not appear on the periodic table. Perhaps, like Frances, they tend to keep this to themselves, unless asked.

The new genetics raises very old issues in a new context. Determinism, for instance: How much control do we really have over who we are? Is it predestined before birth? Or the brain/mind dichotomy: What is the consciousness, the soul? What are we, beyond molecules?

These issues focus our attention on a vast and tantalizing chasm of ignorance about human nature, matters beyond the limits of science. There appears to be no way to address them by testing theories. As always, science pushes at its own boundaries. As always, many people look beyond it for comfort.

Frances was scheduled to return to the clinic a last time, to hear the result of her blood test, on November 6. Soon it became apparent that there was a large chance there might be no result at all.

The geneticists were analyzing six different markers. As it turned out, her father was homozygous for five of them: both chromosomes had the same type, and it would be impossible to use them to judge whether he had passed a healthy chromosome or a flawed one to his oldest daughter. Only the sixth marker, for which he had nonmatching types on his pair of chromosomes, could be informative.

If Frances's mother had the same marker types as her father, it would still be impossible to judge which chromosomes Frances had inherited from whom. Suddenly the odds were 50–50 that she couldn't know at all.

Then Frances changed her personal forecast about the outcome of the test. She decided it would be uninformative. She wouldn't get a positive result; she would get no result at all.

The odds were in favor of that outcome: now there was only one chance in four, 50 percent of 50 percent, that she could learn she was not affected.

Then the first time they tried the analysis of the sixth marker, they couldn't get a reading for technical reasons. Two days before the scheduled appointment, Frances got a call saying they would have to put it off for a week. "This has never happened twice to anybody," the counselor said. "We don't expect it will." But the test failed a second time.

Technical difficulties are one irritation; another is what genetic counselors refer to as the difficulty of establishing "phase." Many genetic markers, including the best markers for Huntington's disease known in 1987, have more than one variant. They can be inherited as any one of a number of genetic varieties, or *haplotypes*. In other words, a restriction enzyme may produce several different-size fragments from a particular chromosomal region, in a particular family.

Think of eye color and hair color: There are a number of basic varieties, but any individual in a family can have only one eye color and one hair color. In the same way, each person in a Huntington's disease family will inherit only two different sets of fragments in the region of the disease gene—one for each member of his or her pair of chromosomes—but throughout the family there may be several different varieties. Let's call these A, B, C, D, and E.

Imagine that you are a person at risk of Huntington's disease, which, as in Frances's family, travels down your father's side of the family tree. Your mother's family is unaffected. Fortunately, genetic counselors tell you that your family is "informative": Both of your parents are still alive, and you have enough other surviving relatives with Huntington's to allow them to analyze the genetic pattern for the disease, if all goes well. They will be able to determine some relatives' genetic types by analysis of their blood cells, and in other cases they will be able to state what genetic type a relative has by deduction, given the genetic types of his or her children or parents.

Imagine that studies show that, in your family, no one with the D or E haplotype has ever developed Huntington's disease; in fact, the E haplotype travels only on your mother's side of the family. These two haplotypes can be eliminated from consideration, although they still help to sort out patterns of inheritance for the other alleles. People with Huntington's disease may have the AB type, the BC type, or the BB type. The AC type is never inherited with the disease in your family. It's obvious that, in your family, the B allele carries the Huntington's gene—but perhaps not always.

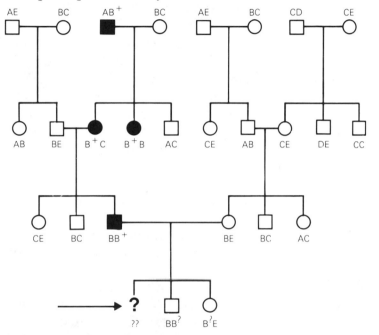

An insoluble phase problem. Because both their mother and their father are able to contribute a normal B-type gene to any of the children in the latest generation, it is impossible using this marker to determine whether a particular child who possesses the B type will develop the disease.

Not always, however. Some tall, dark strangers are handsome; some are not. Similarly, some people with the B allele, including your mother and any of her blood relatives who have it, are free of Huntington's disease. The problem, for you or your siblings or any of your paternal-side cousins, is to determine whether or not a B allele they have inherited is one that carries the disease gene. That is the problem of phase.

If B were the only good marker for Huntington's disease in your family, and if you happen to have inherited a type B allele, matters could become difficult. If your father was a BB homozygote (so that he could have passed you either a normal-type B allele or an affected one)—or, alternatively, if you and both your parents have the AB type—you're out of luck, in one sense. Genetic testing, linkage analysis, cannot determine whether or not you have inherited your father's B allele. Until the discovery of the gene itself, which would lead to precise molecular tests, your odds of developing the disease at some time in the future would remain, as far as anyone could tell, 50–50. You'd keep waiting.

Until a test exists for the gene itself, predictive testing for Huntington's disease will always be a matter of odds—the likelihood that a recombination between two chromosomes has not severed the disease gene from the relevant marker when the person undergoing testing was conceived. Also, genetic tests may never be able to predict the severity and age of onset of Huntington's disease in a particular individual. Certainly the current ones cannot.

Frances learned her future on Thursday, November 19. The appointment was set for 11:00 A.M. Frances awoke at six, with severe pains in her neck and shoulders. She sat up, and her vertebrae cracked. She cried out, lay back for a moment, then got up.

She chose a special dress for the occasion: silk, in vivid blues and reds, with a frill at the hem. A dress you could dance in. Slow motion, she and Jeff got dressed, ate breakfast with the news on the kitchen television, got ready to leave. The au pair played with the children.

Then they drove into the city. Neither said much. It was a stunning day for late autumn, as vivid as Frances's dress, and mild, like spring. There wasn't much traffic. They arrived early.

Frances and Jeff killed time in a coffee shop. Afterward, they wandered around the city sidewalks, with twenty minutes left to wait. And, the thought occurred to Frances, twenty minutes left to hope. She said the thought aloud. "There's twenty minutes left to hope," and, later, "ten minutes left to hope. This is the last time. We're going to know the answer."

At five minutes before eleven, they entered the hospital lobby. A few yards away, they saw the doctor enter by another door. Jeff took note of his demeanor, and made an observation he kept to himself. Frances wanted to hide from the doctor in the lobby, but he noticed her. He passed on.

They and the doctor arrived separately at the room where they were to meet. He showed them into a small chamber with a desk and two chairs. Frances sat down, but before Jeff had time to take the other seat the doctor told them. "Everything's fine. It's fine."

"You're joking," Frances said. "I don't believe it."

She didn't cry. I'm a very controlled person, she admitted as she recalled the moment. Perhaps too controlled. She had prepared herself, one moment weeks earlier, and now she held back.

Then the joyous words began to gush out, and the doctors beamed—a soothing moment for people who sometimes have to impart the worst news instead. Frances and Jeff stayed in the clinic, exulting, for about a half hour, and then went down to the hospital lobby to find a telephone.

The first person they called was Frances's father, in the nursing home. "He can't talk all that well; he's pretty weak now," Frances told me. "He's just not there as much; he's like the shadow of a person. There's no vividness. But he told me he was very happy, that this was what he thought everyone should do."

Next they called her mother. The reaction was not exactly what Frances expected. "I'm so pleased for you, it's fantastic," her mother said. "This reminds me of other moments in my life when I've had similar things happen." As Frances got off the phone she wondered, What does she mean? When in her life has anything exactly like this happened?

"What she said was inappropriate," Jeff agreed, "but probably understandable. She still has other children and grandchildren to worry about."

Frances had not told her sisters she was going through with testing, so she decided to inform them of the results by letter. Madeleine called as soon as she got the news. "I wish I had been there," she said. "You would have gotten a huge hug. I am so thrilled for you!"

With Jennifer, the youngest sister, it was different. Her husband called Frances after the letter arrived. He was candid: Jennifer had been devastated by the news, he said. She couldn't do anything all day. She cried, she called her mother.

A few days later, Frances had to call Jennifer about Christmas plans. As soon as Jennifer heard who was calling, she said, "Oh! Congratulations."

"It was a fairly lukewarm greeting," Frances told me. "And that was all." They have not discussed it since.

Since the beginning of history, Nancy Wexler often points out, people have tried to predict the future. In the past, people read tea leaves, entrails, crystal balls, cards, stars. Today, we read DNA. The Greeks would not act on any important matter without consulting the Oracle of Delphi, she reminds us. Macbeth sought the prophecies of witches. Often, people misunderstood the prophecies, or failed to use their foreknowledge wisely.

"The critical question placed most often before the oracles, ancient and modern," she has written, "is, at some elemental level: Am I going to live or die on this mission? When and how will I die? Am I going to suffer? People are mesmerized by the question, and terrified of the answer."

Wexler's mother died of Huntington's disease. When the genetic marker was discovered, Nancy Wexler was in the middle of the high-risk period for the onset of the disease. She chooses not to reveal whether she has had the test herself. But she has become an active force in spurring genetic studies on, and also in helping people to contemplate their meaning.

The head of one Huntington's disease clinic said that when many people think of Huntington's disease today, they no longer remember Woody Guthrie. They envision Nancy Wexler: Tall, striking, outspoken, vigorous, at risk.

In a way, that's unfortunate. Nancy Wexler is not Huntington's disease. Neither is Frances. Nor was her father. They are both, like everyone else, people dealing with what life presents them. How they deal with it is who they are. The tragedy of Huntington's disease is that it leaves someone like Frances wondering whether it was her father who was irascible and unfair, or whether to blame the disease—and who he would have been without it.

Nancy Wexler speaks with conviction and wisdom about

the dilemmas presented by modern genetics, and people always listen attentively. Not only is Wexler head of the Hereditary Diseases Foundation, and a valued consultant to many genetic studies and the node around which others have been organized, but her words come from the soul.

The oracle of DNA is not really unprecedented at all, she points out. We have been grappling with the definitions of good and evil knowledge ever since Adam and Eve and the apple. There is no question that the new genetics can bring us great good. But we must be careful to learn how to interpret the oracle correctly, and to use its predictions with wisdom.

In a sense, we are our DNA. In other senses, we are not, and we must not forget that.

Frances had finished her story. It was lunchtime; the children were banging at the double door. The coffee was cold.

"Is that all you had to ask me?" Frances said.

"Unless there's something you'd like to add," I replied.

"There is one thing," she said. "I don't know if you want to include it or not. Curiously enough, dramatic and meaningful as it was, the test result does not make life rosy afterward. Not completely. I say this because it completes the picture."

After the result, to Jeff's complete astonishment, Frances became irritable. She developed some classic symptoms of clinical depression: disorganized thoughts, waking very early in the morning, a lack of motivation. She felt restless, and longed to go back to work.

"Living with being at risk was enough of a challenge that I didn't feel this need," she said. "I didn't need to do much to be a hero, and I was comfortable with that."

Before she got the result, she could find purpose merely in going through with it. Jeff was obviously impressed. "I don't know how you do it," he kept saying. "I would be a basket case. I am a basket case! But you're so cool, so serene." Frances remembered this admiration afterward, and missed it.

"And then it's over," she went on. "That chapter is finished forever. And you're no longer Frances the heroine, facing the challenge of a lifetime. You're Frances the pampered housewife, with a Harvard law school education. What are you going to do now?"

Over a period of months, the depression and restlessness gradually lifted. Frances began doing volunteer work with a church group that helps needy families in her area. She and Jeff began spending more time together than ever before. Together, they took up golf and dancing—both new interests that Frances would never have taken up beforehand, for fear that they would only point out that her body was beginning to fail.

Frances still intends to return to the law someday, after her children are in school. But she would defer job hunting for quite a while. As we spoke, although she did not know it then, she was carrying her third child. Life goes on, remarkably ordinary.

11

UNKNOWN TERRITORY

During a coffee break at a genetics conference, I leaned back in a comfortable chair and closed my eyes. A conversation began somewhere behind me, within earshot. I did not open my eyes or stand up and walk away.

"I just called back to the lab," I heard someone say, with a sigh.

"Bad news?"

"No news. I can't believe it. I go away, nothing happens. Except for a second result on a possible Turner's."

(Turner's syndrome is a chromosome imbalance that affects girls. It causes short stature and abnormal development of sex organs. Neither defect is life-threatening or disabling, if you ignore infertility.)

"That's a tough call. How do you counsel?"

"Depends . . . Last one we had was an Oriental couple. That's easier. Orientals have a cultural preference against girls, anyway."

"So what did they do?"

"They aborted."

"You could always counsel them to go ahead, and just tell

her later on that she'd probably be infertile." A long pause. "I don't know what I'd do if it was me."

"Your client, you mean?"

"No, my fetus. Sometimes I wonder."

"Oh, I know what I'd do. No question."

"You'd abort?" There was another pause. "I'm not so sure."

"Well, I don't want to have children anyway."

"Why not?"

"Couldn't ever see much point in it."

I could not help but wonder what the genetic counselor who saw no point in having children had actually said to the Oriental couple.

Genetic counselors are trained for compassion. It is their business to explain the complex issues couples need to understand in order to find their way through some of the most difficult decisions anyone ever has to face. I have the highest regard for the profession, and regret portraying them only through the unguarded words of an anonymous counselor in a private conversation. I do so to illustrate a point: counselors are exactly that—a profession—and like all professionals, they may regard some aspects of their work as routine. Some situations, some cases.

Genetic counselors are not all insensitive; far from it. Neither are they unique in sometimes being so. In *Proceed with Caution*, a book about the ethical and practical implications of gene mapping, pediatrician Neil Holtzman writes that studies show many people go for genetic counseling because a doctor has said they "have to," often without explaining the heart-wrenching outcome that may follow: aborting a wanted fetus.

We have many vivid images of the darker face of science: the sorcerer's apprentice, Dr. Frankenstein's monster. Hiroshima. Chernobyl. Hitler's eugenics. The genome project too raises specters, but they are more subtle and less diabolical (considering the restraints that already control medical science in this country). They are like that faceless counselor I heard during the coffee break, or the doctors Holtzman describes.

Unquestionably, the genome project will give us new information about ourselves. But before it reaches us individually, it will be filtered by any number of medical professionals. It may arrive with their value judgments attached. It may be inaccurate, as any medical result can be. Like a name on a catalogue mailing list, genetic information may even be channeled to places we wish it would not go, perhaps without our knowledge. Above all, we will need to be informed, and vigilant.

H. G. Wells once called science a match that human beings just have to light. In their insatiable curiosity about our molecular machinery, scientists have come up with some very seductive reasons why we should pay them to map and sequence human DNA.

They crusade. They speak of it as a noble cause, as well as a scientific and intellectual adventure, says Robert Cook-Deegan, a bioethics adviser to Congress. Scientists regard the genome initiative as a chance to test their character, he adds, "to practice virtuous acts and conquer temptations."

As always, however, scientists will not be the only ones asked to practice virtue and conquer temptations. They will launch their new knowledge into the world—not a Brave New World, or even a new world at all. The same old world, with the same shams and human foibles, now presented with a new opportunity and a new problem.

The genome project now seems to have a momentum of its own, beyond the impetus from scientists and the biotechnology and pharmaceutical industries. It's astounding how quickly events have progressed.

In 1980, when David Botstein and his coauthors laid out the possibility of mapping the human genome using RFLPs, many people who could understand the new idea found it wild and impractical. Clearly, the question was, Could we map the human genome?

Five years on, at a 1985 symposium at the University of California at Santa Cruz, the debate had evolved. The question was, Shall we map the genome? Describing that conference, journalist Stephen S. Hall portrayed human gene mapping as "preposterously possible."

In the same year, at the dedication of a new laboratory at Cold Spring Harbor, the geneticist Renato Dulbecco proposed mapping the genome as a boon to cancer research. James Watson noticed little reaction from that audience composed mostly of cancer researchers, not human geneticists.

Four years later, none other than the American Medical Association cosponsored (with the Alliance for Aging Research) a conference brightly titled "Unlocking Potential: The Promise of the Genome Initiative." But the tone of the agenda was somber, and urgent: How shall we map the human genome, and what shall we do with the information? And (for almost the first time before a general audience) What hazards do we face? Besides scientists, the speaker list included ethicists, politicians, lawyers, and representatives of biotechnology and insurance companies.

There was no question, by then, that the job should and would be done. In fact, it was already partly done. A low-resolution linkage map of the genome was all but complete. Numerous scientists in several countries, including those in the Cantor-Smith team, were racing to complete physical maps of individual chromosomes.

Both the National Academy of Sciences and the Office of Technology Assessment had come out with major reports in favor of gene mapping. The American government had already committed substantial sums to the idea—almost $50 million for fiscal 1989, shared among several institutes. The latest edition of Victor McKusick's catalogue of genes was 1,752 pages long. Over four thousand human genes had been cloned, linked, or mapped.

The project had evolved from a haphazard assortment of genetic experiments into something loosely called the "genome initiative" (nobody seemed entirely clear about what that was). Many of the necessary elements were already in place: computer data bases, numerous repositories of cells and DNA, various different maps, and a growing number of high-tech machines for analyzing DNA.

At the Washington conference, speakers were debating how to divide up the work. James Watson, by then the appointed head of the National Institutes of Health's genome project, made the half-serious suggestion that the mapping should be parceled out, chromosome by chromosome, internationally. Britain should have "its" chromosome, France should get one, maybe a really big one for the Soviet Union (which had just announced a 40-million-ruble gene-mapping program).

Some speakers expressed great concern that an effort by the European Economic Community to set up its own genome project had been either suspended or halted (it was not clear which), at least partly because of ethical concerns. According to a recent article in the journal *Nature*, some West German politicians had accused the European program of endorsing "racial hygiene." Indeed, its motives did seem to echo the rationale used by Nazi eugenicists. "Genetic diseases are said to be 'distressing' and 'socially expensive,' " the *Nature* article reported, summarizing the justifications for the European project, "so that the possibility of relieving or preventing them is promising."

People on the dais in Washington were visibly relieved that the U.S. program seemed quite vigorous. Politicians and businessmen spoke of it as a way for the United States to grasp a firm international advantage in biotechnology. "Our competitors in this field are dancing a very fast tune," said Senator Pete Domenici, in an address specially videotaped for the meeting.

Absolutely no speaker suggested that the genome should not be mapped, or that the mapping should not proceed, if very carefully, toward finding the exact genetic sequence of the human genes. The reasons in favor of doing it seemed obvious.

Watson mused about them privately, a few weeks later, in his own office at Cold Spring Harbor. "The chief reason society's interested in the subject," he reflected, "is their concern with the prevention of genetic disease—which we now realize is such a large component of disease. The scientists are interested because of the enormous information it's going to give us about life. They're two separate issues."

The genome project is not targeted at genetic diseases per se. That kind of targeted research will proceed as usual, quite apart from the genome project (but benefiting enormously from it). The National Institutes of Health's genome project, Watson said, will focus initially on learning the genomes of simple organisms—bacteria, yeast, and *Drosophila* (fruit flies)—as models of the human genetic system. He had also proposed to support a collaboration with British scientists, who were well on the way to completing a map of the genome of the common roundworm, which is about as large as that of one human chromosome. At that moment, many of his own concerns were bureaucratic: How many scientists it would take to complete the genome, with what skills, in what combinations?

"I think our program will be very cost effective," he added. "We'd like to do the *Drosophila* project for $100 million. We're now probably spending $200 to 300 million on *Drosophila* studies. So your total cost of the project is less." The payoff should be a fuller understanding of the meaning of the genetic sequences that are beginning to emerge from the human genome.

"We'll start with bacteria," he went on. "We're going to see all the genes, and with [them] a lot about the proteins which make possible the life of a single cell. For *Drosophila,* you're going to have all the information to see all the proteins that make possible the existence of a fly. And we're going to build up to all the proteins which are used together to make a human being—which proteins build up our brains, our toenails, the whole business.

"It's an immense objective," he continued. "The understanding of human beings. . . . Instead of saying we're just going to try and understand muscular dystrophy (which we did through the function of a gene), we're now going to say we're going to try to understand life, by learning the genetic message."

Exactly a week later, Watson's de facto counterpart, Charles Cantor, reflected on his role within the Department of Energy's arm of the genome project. The agency had not yet appointed all of the officials for its own genome program; Cantor was the only person to be named head of a DOE gene-mapping laboratory, and he was chairman of the DOE's genome project advisory committee. He said he saw his role as "scientific spokesperson" for the DOE on this subject, and as the glue that held the various DOE projects together.

An important part of the DOE's contribution was to develop the tools for physical mapping of the genome, and to begin to use them. He envisioned an operation like a small biotechnology company, or a factory, where much of the work is automated and employees are given small incentives for superior production levels.

Cantor logged onto his computer, and printed out a slide that outlined his own personal production schedule. One to five years, it reads: 1 percent of the genome sequenced, 50 percent of the genome mapped. Six to ten years: 10 percent of the genome sequenced, all of it mapped. Eleven to fifteen years (by the year 2000 or at the latest 2005): all of the genome sequenced; all of the genes found.

It's an ambitious schedule, I offered. Yes, but not unrealistic, he replied.

"I've never tried to figure out how much time it will all take," he went on, "but I think . . . a sequence map could be done in a year or two at high resolution." At that very moment, for the first time, he launched into the calculation.

"Let's say a twenty-kilobase map, which is all anybody thinks of right now, of chromosome 21. Let's see." The wheels were turning in his head again. "That means twenty-five hundred sites, each of which requires four hundred base pairs of deep sequence. . . ." His voice trailed off. ". . . One automated sequencer can generate ten kilobases of raw sequence a day. That's no sweat. So one or two hundred days of sequence with one sequencer to put such a map together, if we have correctly automated all the steps."

He went on, without pausing. "Now that I've done this calculation, which I've never done before, it sounds like what I said before: It's hardly pushing."

"We hope to complete a restriction map of chromosome 21 within a year, using the older strategy [the kind of project Akihiko Saito labored at]," he had said earlier, "because most of the lab is devoted to it, and because it's working. The new PCR strategy, if it works as planned, makes the job . . . fairly trivial." It eliminates the need to create libraries of genes and link them up, he said, or to make data bases or carry out complicated calculations. "That's why it's revolutionary," he said. "That's what I like so much about it."

A few months after this conversation, Cantor and three other eminent geneticists (David Botstein of the biotechnology firm Genentech, Leroy Hood of Caltech, and Maynard Olson of Washington University) coauthored an article in the journal *Science*, in which they proposed a new, universal language for gene mapping, based on the genetic alphabet itself. A set of PCR primers, each one representing a unique small segment of known DNA sequence in the human genome, should be used as landmarks to replace the thousands of probes behind the current genetic map, they urged.

Almost any lab, large or small, they pointed out, now has the capacity to create specific small DNA sequences. PCR is almost a standard genetic technique. New landmarks based on primers could unify research all over the world.

The proposal had two strong arguments in its favor, the geneticists said. PCR primers could not be hoarded and they would not need to be stored in some central repository. The primer language would also eliminate the confusion that arose when everyone had his own alphanumeric label for his own probes of uncertain genetic content.

The idea wasn't really Botstein's or Cantor's or anyone's in particular, Norton Zinder of Rockefeller University told me. As chairman of the advisory committee to the American human genome project, he heard it emerge over the course of several meetings. It seemed so obvious that everyone involved agreed to it immediately, he added; the four authors wrote the paper merely to get the idea into circulation. All that remains to be seen is how quickly it is adopted and implemented.

"Right now," Cantor had told me, "we're still setting the methods. But I think as the methods work better, the rate of production will increase tremendously and it will be satisfying. If we're sequencing ten thousand base pairs a day and analyzing them, I think day by day we will see interesting things. You'll compare it with other sequences. Things will pop out at you. It will be terrific fun."

For pure scientists, obviously, the genome project promises the most gratifying kinds of professional challenge. For people with hereditary diseases and the doctors who treat them, it beckons with hope. For the rest of us, the promises of genetic science are less self-evident, and may also conceal threats.

The recent history of forensic genetics is a cautionary tale. DNA "fingerprinting"—analyzing the restriction-fragment patterns from the chromosomes of criminal suspects and their victims—has become increasingly common, now used as evidence in thousands of cases in the United States. It is often touted as fail-safe, as irrefutable, because any person's DNA blot is even more unique than a fingerprint, if you analyze a large number of probes.

In May 1989, however, four eminent scientists asked the National Academy of Sciences to establish standards for DNA fingerprinting, claiming that much of what had been accepted into evidence represented shoddy science. In one well-publicized murder case in New York later that summer, a judge threw out DNA fingerprint evidence that appeared to link an accused man with a murder victim, based on analysis of a small bloodstain on his watch. The company that performed the DNA analysis admitted, among other errors, that it had used human beings to "eyeball" the DNA blots rather than applying objective measurements, as it had claimed.

A new generation will grow up with genetic tests, one scientist has predicted, and will probably think no more of them than we think of flying. But flying is a good example. We may take it for granted, but no one is completely comfortable with it—what with near-misses, aging aircraft, and terrorism.

We already live in an age of risks. By daring to predict our medical fate, the new genetic technology has the potential to add a more concrete and far more personal element to that deep-seated paranoia you can discover for yourself if you mention pesticides to your next-door neighbor. Meanwhile, there's no guarantee the genome project will provide a quick or satisfying solution to whatever disease may be in your genome.

"What is the truth about the genome initiative?" asked Dr. Robert F. Murray, Jr., the head of medical genetics at Howard University Medical Center, at the Washington conference. "We will have this [new genetic] knowledge, but whether in fact we will unravel mysteries is questionable." Scientists *believe* they will do so, he said. They don't know it for sure.

The history of the first molecular disease is instructive in this regard. For forty years, doctors have known exactly what's wrong with sickled hemoglobin, and now we know it down to the level of atoms. Effective prenatal testing has existed for more than a decade. If there was ever a health problem the new genetics could solve, this should have been it.

Yet, according to a 1987 survey, less than 5 percent of expectant couples at risk of producing a child with sickle-cell anemia opt to have prenatal diagnosis for the condition. In that year there were 272 prenatal tests for sickle-cell anemia in the United States. About 6,500 couples were at risk.

In 1974, Dr. Murray stood up at a meeting and predicted that within a decade there would be a cure for sickle-cell anemia, based on the knowledge of its cause. "We've made some advances," he said at the Washington conference fifteen years later, "but they have nothing to do with genetics. If I had a different complexion, my face would be red."

For all the molecular knowledge about sickle-cell anemia, there is still no specific therapy. The best treatments are still merely supportive: Painkillers for the crisis, dialysis for kidney failure, and (probably most important) preventive penicillin treatments to ward off fatal infections among infants with the disease.

In the 1950s and 1960s, armed with molecular knowledge about sickled hemoglobin, scientists took off after the disease with a vengeance. They tried to create drugs that would trap the defective sickled hemoglobin, to prevent it from polymerizing and causing sickle crises. The new drugs seized cell membranes and other biological molecules just as easily; they caused more problems than they solved. Drugs that disrupted the sickled hemoglobin polymer also had toxic side effects.

"A lot of people got involved, and then came up against double-edged swords," said Griffin Rodgers, who is doing sickle-cell research at the National Institutes of Health. "You attempt to do something positive. It works quite well in the test tube. Then the patients get dehydrated, or develop neurological problems or cataracts. The bright and promising investigators say, 'I've worked on this for ten years. Let me move on to other areas.'"

Within the next year, an article in *The New York Times* would proclaim a "turning point" in sickle-cell research. A number of scientists were quoted as showing a renewed interest in looking for new treatments. The greatest excitement, it said, surrounded the possibility, due to genetic advances, of creating a new animal model for drug testing by inserting the sickle-cell gene into laboratory mice. Still, all researchers had to offer at the time were hopes and promises.

If the situation was so unfavorable, why did so few couples

grasp the opportunity to detect and abort an affected fetus? There may be important explanations in the way prenatal care is provided to many blacks in the late twentieth century. But there's another possibility. A quarter century after the molecules of sickle-cell anemia were an open book, genetic counselors could not offer a couple very useful information.

Some 80 percent of people who have sickle-cell anemia—not the trait, but the full-blown disease, with two copies of the sickled hemoglobin gene—are not seriously debilitated. Given good medical care, most of them live well into adulthood. The crises are neither frequent nor terribly painful; most of them are not much worse than arthritis, and can be relieved with Tylenol or other over-the-counter painkillers.

For the remaining one in five people, sickle-cell anemia is truly a tragedy. They have crises every few months, and require strong painkillers. Much of their lives, they are in severe pain. They are prone to grave complications such as kidney disease, and are in serious danger of early death.

Doctors have no way to predict, before birth or after, which category of "affected" a child will fall into. Consider what two carriers of the sickle-cell gene hear from a well-informed genetic counselor, before prenatal testing: There's a 25 percent chance that your fetus has the disease, but an 80 percent chance that it will live a fairly normal life anyway. We can tell you with absolute certainty whether or not your fetus is affected (because we know the gene); the 80 percent figure is a complete gamble, because we don't know how to account for it. Would you abort a fetus on such a basis? Would you even take the test?

How many members of the general population will take the opportunity to learn whether they carry the newfound cystic fibrosis gene remains to be seen. For what it's worth, a study at a Connecticut children's hospital found that only 8 percent of couples offered prenatal testing for muscular dystrophy took advantage of it. All of those were members of families in which one child had already been born with the disease.

At present, most prenatal genetic tests are done to rule out gross chromosomal abnormalities: sporadic major accidents of reproduction such as Down's syndrome, the inheritance of an extra chromosome 21. Many couples who have prenatal testing do so because they know they are at risk. Having had one child with a rare hereditary disease, they seek to learn whether additional children will be affected. It is these people for whom the gene mapping has done the most to date, by increasing the accuracy of the predictions they hear from genetic counselors.

If all goes well for gene mapping, doctors will someday offer us the opportunity to test our fetuses for susceptibility to less dramatic and debilitating diseases, conditions such as Alzheimer's disease, high levels of blood cholesterol, cancers that emerge in late adulthood, depression—many of which would allow an affected person many years of meaningful life.

The new knowledge from gene mapping offers parents the opportunity to forestall such problems in their children's future. It also threatens to leave prospective parents with the difficult problem of deciding whether to eradicate a fetus with such a future or to bring it to life, gambling that science will solve the medical problem during its lifetime.

Will parents cloud such a person's childhood by being overly vigilant? Or will the person the fetus later becomes—or someone responsible for that person at a later date—choose to bring a "wrongful birth" lawsuit against the parents? Such things have happened already.

Several prominent scientists have argued that people have an ethical obligation to destroy fetuses they know to have serious birth defects. None other than Linus Pauling said that people who do not do so "add to the amount of human suffering [and] should feel guilt for their actions."

More widespread and sophisticated prenatal testing could turn the "right to life" issue upside down, if women had to assert a right to opt *against* abortion. More likely, public attitudes about the handicapped will shift.

Many couples already seem to expect to deliver the "optimum baby." Will their perceptions of "optimum" shift to even trivial matters such as poor vision or body stature? Will parents want to treat or abort a male fetus destined to be short? Poor vision is also a handicap, after all. So is lack of fur, for that matter. We ignore shortness, and we solve poor vision and bare skin with simple aids: glasses and clothing.

Support groups and lobbies for people affected by rare diseases are understandably uncomfortable at promoting abortion as a "solution" even for grave disorders, because it implies the nonexistence of people they love. It also may weaken enthusiasm for other solutions.

In our children's generation, people may be far more critical of the frankly handicapped and more likely to view them as "mistakes," as the doctor and philosopher Leon Kass puts it, as people "who would not have been, if only someone had gotten to [them] in time." A high regard for the valiant individual who surmounts a handicap or achieves great things despite it—Sylvia Plath, Vincent van Gogh, the physicist Stephen Hawking, or the people who win at Special Olympics—may be seen as an antiquated and coy attitude as the reasons for valiance disappear, along with the people who have the handicaps, who may never be born at all.

It may become more common to view people's physical or emotional misfortunes as their fault, if we can be sure they arrived here without serious inborn flaws. On the other hand, the ability to know a person's genetic makeup raises anew some very old philosophical questions summed up by the word *determinism*. Are we born predestined to a certain future? How much freedom do we really have?

"I frankly think that the public is not so stupid as to buy into genetic determinism again," Cook-Deegan wrote in a letter to me. He added that most people assimilate more scientific knowledge than they get credit for and predicted that the new knowledge from genome research "will all conspire to give us a less deterministic view of ourselves, just as physics has moved away from rigid mechanics to probabilistic predictions. Genome research should foment this more complex view of humankind, not reinforce the simpleminded Mendelian views that so many fear."

Many individuals, indeed, are not simpleminded. Institutions and organizations often have to be, in order to function.

Consider the matter of health insurance. It's a business of risk assessment. What happens when the true risks are known?

Even today, when it comes down to it, only healthy people can get health insurance—and some 37 million Americans cannot. If someone who applies is already ill or at known risk, insurance will either be refused or made prohibitively expensive.

Consider that you are about to go to the doctor for a predictive genetic test. The tests that exist today are quite expensive, in the thousands of dollars for Huntington's disease, to choose one example. But you may not want to ask your insurance company to pay for this test; it will ask to learn the answer too. What kind of conditions should insurance companies be allowed to set if predictive testing becomes widespread?

Again, the history of sickle-cell anemia is a cautionary tale. In 1975, about 10 percent of insurance companies either denied insurance to sickle-trait carriers, or raised their rates. Finally, some well-meaning black legislators tried to introduce widespread screening in schools and before marriage. Some people began to regard sickle-cell testing as a veiled attempt at genocide. In the end, genetic testing for the sickle-cell gene drowned in a sea of negativism.

"Does a genetic predisposition equal a medical history?" asked geneticist C. Thomas Caskey at the Washington meeting. "There has been no public debate about this. If such were the case, you can imagine there would be tremendous inhibition of the use of this technology."

Questions for and about the insurance industry spiral on and down until they are out of view. If an insurance company is asked to pay for a prenatal test for a genetic condition, should it be allowed to say that if the fetus is affected and the parents choose not to abort, it will not pay for the child's medical care after birth? If it does not, who will? Is the future held out by gene mapping in fact compatible with a private health-care system such as we have in this country today?

If not, should society be forced to pay for the health care of children given life despite prior knowledge of their condition? Whether or not we come to terms with them, the questions exist and will affect our society.

At the moment, 80 percent of all large employers are self-insured, and health insurance is a major and growing cost of doing business. By the mid-1980s, several companies were developing blood tests that companies could use to assess employees, or prospective employees, for their risk of future disease. This would give an employer the attractive option of hiring only people who would be good insurance risks. Already, some 6 percent of employers are refusing to hire smokers for health insurance reasons.

The situation paints a scenario for the future: Someone takes a blood test while applying for a job. The blood test shows him to be at risk of a disease, a risk he is unaware of. The employer simply declines to hire him. He may be none the wiser; he may also be unemployed.

There is a great difference between a predisposition or a potential risk, which can be pointed out by genetic tests, and the actual existence of a condition. (Think of high blood pressure and heart disease, for instance.) But employers and insurers may not bother about the distinction if there is no easy treatment for the threatened disease.

Perhaps wealthy Americans in the future will buy the luxury of informed choices about their future, by paying for their own genetic tests, and the poor, as always, will just get by. But not necessarily forever.

For an uncertain period of years or decades to come, predictive testing and all the other offshoots of gene mapping will be halfway technologies, to borrow the term used by the physician-philosopher Lewis Thomas. Like the iron lung once used for polio victims, they will be imperfect and incomplete answers, stopgap solutions awaiting a better approach.

The pharmaceutical industry is looking forward to crafting drugs tailored as perfectly to a molecular disease as a key is fitted to a lock. It also expects gene mapping to give clues to which people are subject to which side effects, to elevate drug treatments from the current status of "herd therapeutics."

In the near future, it should become possible to predict diseases, to avert them, perhaps even to eradicate them—if not by aborting every fetus, then by altering the gene that causes the disease inside the person who has it. Gene therapy. It's being rapidly perfected in animals, and now it has begun to be tried on human beings too.

In early 1989, W. French Anderson and a team at the National Institutes of Health gained permission for the first authorized experiment to inject genetically engineered cells into human beings. The studies, in which people were injected with human cells containing a detectable marker in order to monitor cancer therapy, involved only individuals almost at death's door. The gene alteration was not directed at treating the cancer, only at following the progress of traditional treatments and, incidentally, studying the effects of the gene alteration.

Actual attempts at human gene therapy were not far behind. A year later, Anderson's team gained the first preliminary approval from NIH for a proposal to test a form of actual gene therapy. It is aimed at correcting a severe inherited immune deficiency in children, by directly altering genes in their blood cells.

"Few discussions of gene therapy at scientific meetings and in publications still argue its need or potential place in medicine or its ethical applicability," wrote Theodore Friedman, a pediatrician and molecular geneticist at the University of California at San Diego, in June 1989. "[R]ather they emphasize technical questions." Gene therapy was beginning to be touted not just as a treatment for rare blood and immune disorders but for common conditions such as Alzheimer's disease and emphysema—if the relevant genes could be identified.

The details and ramifications of gene therapy would be ample material for a whole other book. Suffice it to say that there seems to be a general ethic in the medical research field that we will not act to alter the germ lines of human beings, to create changes in the reproductive cells so that an intentional genetic alteration will be passed down from generation to generation. "This is an exciting area not applicable to man," Caskey said.

What genetic engineers like Caskey contemplate at the moment would stop at changing the nonreproductive body cells of an individual, such as blood cells or organ cells, a step that is fundamentally no different from, say, an organ transplant. Even so, gene therapy remains controversial, and an ethics board responsible for approving all such experiments in the United States is moving very carefully.

One question it does raise for us—but, on the other hand, so does gene mapping in itself—is to what extent, and how, we should engineer ourselves. In a sense we have been doing so all along, as we rescue ever more children with deleterious genes so that they survive and pass along genes for things like hemophilia and heart defects. But the issue is subtly different when it involves not saving lives but "preventing" ones destined to develop any one of an ever wider range of abnormalities.

Will gene mapping ultimately lead us all, one by one, to do to ourselves and our offspring what the eugenicists wanted to do for us: to characterize our selves as our genes? Once we begin to identify the genes behind traits, not just behind diseases, will we begin to classify genes as "good" and "bad"? Will we be wise in making the distinction? Do we know how?

What's more, will we begin to focus so single-mindedly on heredity that we neglect even more than we do already the influence of nongenetic factors such as poverty, broken homes, inadequate school systems, racial prejudice, toxic chemicals, social isolation?

James Watson has waking nightmares of his own. One of several imaginary scenarios he invoked in a private conversation was the following: another J. Edgar Hoover, with access to a centralized data base of genetic profiles of private citizens and public figures. (What critic of the president has a gene for schizophrenia?)

One eminent science historian said of Watson's codiscovery of DNA's structure that "there has hardly been a more decisive breakthrough in the whole history of biology." Yet nearly a half century later, as head of the U.S. genome project and also as a father, he is troubled by some of the implications.

"I think about the ethical issues a lot," he told me. "There are a lot of them. Should you have a legal right to demand someone accused of rape to give a DNA sample? Should you, if you're running for president, have to say what your DNA constitution is?"

At press conferences, Watson can be glib and brash. Reporters sometimes have to protect him from his own ill-considered quotes. But he sounded quite sober that day. He began by alluding to the human cost of our current genetic ignorance. He spoke of his lawyer's child, who has muscular dystrophy, and of another person he knows who has cystic fibrosis. Watson spoke eagerly of the impending discovery of the CF gene (it had already happened, though neither of us knew it), because it would allow couples to opt to learn before a pregnancy whether or not they might someday have to deal with the disease. At the same time, he said, he feels strongly that adults should not be able to carry out complete genetic screening tests on children after their birth.

"Schizophrenia clearly has a genetic component. To the extent that it's a genetic affliction, you can't escape from it," he pointed out. "I can see [a time] fifteen years from now [when] we can look at an individual and see how healthy his antioncogenes are [whether or not he is genetically protected against cancer]. What do we want to do with this information? What is the effect it's going to have? Do you want to be scared for the rest of your life? Do you want the knowledge, or do you not want the knowledge?''

He also feels that genetic information about individuals should never be data-based in the way credit ratings are. Watson clearly feels that those involved in the new genetic research are responsible for ensuring that the new knowledge is used prudently.

Before the Nazis began to exterminate Jews, he pointed out to me, they began with schizophrenics. "The lists were drawn up by the academics," he said. "And in Germany, the professors of genetics kept their jobs [after the war]." Eugenics and its abuses "came out of a respected scientific tradition; you couldn't trust the academics."

This time, he is resolved, it will be different. One of Watson's earliest steps as director of the genome project was to set aside 3 percent of the budget for ethical studies. "I want to proceed with the program, but I think we also have to be careful that we're not also going to harm people," he went on. "That shouldn't be on our conscience."

Like every previous generation of humans, we leave our children a legacy of the blood, written in their genes. We will be the first to leave them a profoundly new legacy, a written catalogue of that genetic information. They will also inherit the burden and the challenge of learning for themselves what it means, and deciding how to use the knowledge.

Perfectly timing their course to mark the end of the current millennium, geneticists have embarked on a great new journey of discovery. The outcome will be at least as momentous as that of another expedition that took place in the middle of this millennium, five hundred years ago, when Christopher Columbus set out over the ocean, curious to see the other shore.

Think of the upheavals afterward: The world expanded from flat to round. Europe awoke to the existence of a new world, and to new species and new cultures. Great European empires rose. Great South American empires collapsed. Whole nations of Indians eventually vanished.

Columbus set off in search of treasure. His followers did bring back some treasure from the New World, but much less than they hoped. Meanwhile they discovered many intimate surprises, great and small: tobacco, for instance; syphilis; popcorn.

Like Columbus leaving the coast of Spain, the explorers of the human genome cannot yet see beyond the horizon. They can only imagine what lies there, and dream about it. The journey is a little frightening, but also very exhilarating. They have persuaded us to pay for it, and they are on their way. There is no going back now.

The rest of us are like Ferdinand and Isabella. We are paying for the voyage, and we know it is under way. But we are left behind on terra firma. All we can do now is wait to see what the explorers bring back to us.

GLOSSARY

Scientists and students will notice that a number of important technical terms and concepts, such as ribonucleic acid, do not appear in this book. These are not oversights. They are compromises necessary to convey an extremely complex subject to readers who are not scientists.

Alleles: Alternative forms of a gene or genes. Alleles are inherited separately from each parent, as the chromosomes from sperm and egg pair up in reproduction. A person might inherit an allele for black hair from her father, and one for blond hair from her mother.

Amino acids: The 20 different molecules that are the building blocks of proteins. They are strung together in a precise linear order, specified by information contained in the genetic code of DNA, to create the many proteins found in living organisms.

Antibody: A complex protein molecule in the blood, created by the immune system in response to the presence of an antigen (a molecule that provokes an immune reaction). As a key fits a lock, each antibody precisely fits a particular antigen, and leads to its destruction.

301

Autosome: A chromosome other than a sex chromosome; one not involved in the determination of gender. A normal human being has 22 pairs of autosomes and 2 sex chromosomes.

Bacteria: A group of microscopic one-celled organisms that possess outer membranes but lack a complete nucleus.

Bacteriophage: A virus that multiplies inside bacteria. Molecular geneticists often use the bacteriophage lambda to transfer foreign genetic information, such as fragments of human DNA, into bacteria.

Base: One of the nitrogen-containing molecules that are the components or molecular symbols of the genetic code; also called *nucleotides.* DNA contains four bases: adenine, cytosine, guanine, and thymine.

Base pair: Two corresponding or complementary bases inside an intact DNA molecule, held together by weak chemical bonds. The DNA double helix is like a microscopic twisted ladder, in which each rung is made of a pair of nucleotide bases bonded to each other. Normally in human beings adenine will bond only to thymine and cytosine will bond only to guanine.

Cell: The smallest component of life, a complex collection of molecules bounded by a membrane. With a few exceptions, such as red blood cells, all cells are able to regulate and reproduce themselves.

Chromosome: One of the threadlike structures in the nucleus of a cell that contain the information of heredity. Genes are arranged in linear order along the chromosomes, which are made of DNA in association with proteins.

Clone: A group of genetically identical cells descended from a common ancestor cell. All cells in a clone have precisely the same DNA. In nature, clones are produced by nonsexual processes in bacteria and other microorganisms. Molecular geneticists also make clones using a variety of procedures to manipulate DNA.

Complementary: In genetics, a DNA base or a DNA sequence that corresponds precisely to another base or sequence, according to the DNA base-pairing rules. For example, in the DNA double helix the sequence ACT in one DNA strand is complementary to the sequence TGA in the other. (See *Base pair.*)

Cosmid: A cloning vector used by molecular geneticists to package a chosen DNA strand into an infective virus particle.

Crossing over: A normal genetic event that occurs during the creation of sperm and eggs; a transfer of material (most often an even exchange) between two corresponding chromosomes before one member of the chromosome pair is incorporated into a sperm or an egg. The process results in a reshuffling of genes and is useful to geneticists in estimating how closely two genes lie together on a chromosome. If there were a crossover in the X chromosome a woman inherited from her mother, she might possess mostly her mother's X chromosome, but partly her father's X.

DNA: Deoxyribonucleic acid, the molecule that encodes hereditary information. Every inherited characteristic originates from the code contained in DNA, which is a double-stranded molecule held together by weak bonds between pairs of nucleotides.

DNA probes: Small segments of single-strand DNA that are treated so they will be detectable (usually by making them radioactive). Probes are used to pinpoint complementary DNA sequences in the process of *Hybridization.*

DNA sequence: The linear order of base pairs along a DNA molecule.

Dominant: A term applied to a gene that manifests itself when present in a single copy. Although geneticists have begun to discover some exceptions to this rule, in classical genetics a dominant condition is one in which it is impossible to see an obvious difference between a heterozygote, possessing only one copy of a certain allele, and a homozygote, possessing two copies of that allele.

Double helix: The "twisted ladder" structure of DNA, with complementary strands twining around each other.

Electrophoresis: A technique for separating large molecules such as DNA or proteins, by exposing a solution containing the molecules to an electric field. Each different molecule travels at a different rate, depending on its physical properties.

Enzyme: A protein that speeds the rate of a biochemical reaction without altering the nature of the reaction and without being altered itself. Enzymes are the biological catalysts that allow the chemical reactions of the human body, for instance, to take place at body temperature.

Eugenics: The attempt to improve a population through selective breeding.

Expression: In genetics, the way in which the information contained in a gene is converted into the structures and processes operating in a cell. On a larger scale, the observable outcome of the inheritance of a certain gene by a human being.

Founder effect: In an isolated population, the deviation of gene frequencies from the norm in the rest of the species, due to the presence or absence of particular genes in the founding ancestors of the community. The deviation from the norm may be enhanced by various environmental factors acting on the isolated population.

Gene: The fundamental unit of heredity. A gene is a stretch of DNA that encodes the creation of a specific protein or carries out a specific function.

Gene frequency: In the genetics of an entire population, the relative abundance of a particular gene or allele; strictly speaking, the actual occurrence of an allele (a particular variant of a gene) as measured against the theoretical number of times it might occur in the population if every individual possessed that allele at every possible locus.

Gene mapping: Determining the ordered relationships and distances between different genes or DNA segments on a chromosome.

Genetic code: The biochemical basis of heredity; the relationship by which a cell uses the triplets of nucleotide bases ordered along a DNA strand to create the linear sequence of amino acids in a particular protein.

Genetic drift: Random fluctuations of gene frequency in small populations due to the preservation or elimination of specific genes purely by chance.

Genome: All the genetic material contained in the chromosomes of a particular individual.

Genotype: The genetic makeup of an individual; a term often used in opposition to *phenotype*, the observable features of an individual organism that are the result of the genotype.

Heterozygous: Possessing different alleles at a certain locus.

Homozygous: Possessing two identical alleles at a specific locus; having inherited the same variants of a particular gene from each parent.

Hybridization: The process of joining two complementary DNA strands to form a double-stranded molecule. A complete account appears on pages 163–65.

Inbreeding: The mating of closely related individuals.

Library: In genetics, a collection of clones of DNA segments.

Linkage: The proximity between genes or markers, or both, along a chromosome. The closer the markers are, the less likely they are to be separated during reproduction. (See *Crossing over.*)

Linkage mapping: Determining the relationship between DNA segments by studying the patterns of their inheritance in a population. By repeatedly analyzing the likelihood that the relationship between markers and genes is explained by linkage and not by chance, and comparing these relationships for a large number of genes and markers, geneticists can deduce the likely order of genes and markers along a chromosome.

Locus: The specific position on a chromosome of a DNA segment or gene.

LOD score: A number representing the *likelihood of disequilibrium,* for any two genetic markers, the likelihood that they are not in fact located very close to each other on the same chromosome. Because it is easier to think in small numbers than in astronomical figures, the LOD score represents a power of ten. If the odds are a thousand (10^3) to one in favor of two markers being linked on the same chromosome (or one in a thousand that they are *not* linked), the LOD score is 3. If they are a million (10^6) to one, the LOD score is 6. A larger LOD score implies a closer genetic (and physical) linkage of two markers.

Marker: An identifiable physical location on a chromosome, such as a gene or a RFLP (restriction-fragment-length polymorphism), whose inheritance can be traced in a pedigree.

Mutation: Any change that alters the sequence of nucleotide bases on a DNA strand.

Nucleotide: A DNA subunit, consisting of a nitrogen-containing base (adenine, cytosine, guanine, or thymine) chemically linked to a phosphate molecule and a sugar molecule. Thousands of nucleotides linked to each other form a DNA molecule.

Nucleus: The control center of a cell, containing the chromosomes and bound by an internal membrane to separate it from the rest of the components of the cell. Bacteria and some algae do not have nuclei; their chromosomes are not contained inside a special wall.

Oncogene: A gene associated with the emergence of cancer.

Pedigree: A chart showing inheritance in a particular family.

Penetrance: The proportion of organisms possessing a genotype for a certain trait who actually show the phenotype, the observable trait. Reduced penetrance may result from a number of external factors including age, gender, and environmental influences.

Phenotype: The observable properties of an organism. Compare with *Genotype.*

Physical map: A map specifying the locations of identifiable landmarks on the chromosomes, deriving from molecular genetic studies rather than from studies of inheritance. Compare with *Linkage map.* The crudest physical map is the banding pattern of chromosomes that can be observed under a microscope. The finest-detail physical map is a complete DNA sequence, base by base.

Plasmid: A circular piece of DNA found outside a cell's nucleus and capable of reproducing itself independently of the rest of the cell's DNA. Small and relatively simple, plasmids are often used by molecular biologists as vectors with which to insert small pieces of DNA into bacteria.

Polygenic: A condition arising as a result of an interaction between two or more genes.

Polymorphism: The persistence of a certain type of genetic variation that is too great to be explained by random recurrences of the same mutation in different individuals. Polymorphisms may persist because the variants are advantageous under certain circumstances, or because heterozygotes are superior to homozygotes. The ABO blood types are polymorphisms with no obvious advantage; one polymorphism in the hemoglobin gene, sickle-cell anemia, persists in some populations because it confers a resistance to malaria.

Protein: A large molecule consisting of chains of amino acids in a specific sequence encoded by nucleotides in the corresponding gene. The structural components of cells are made of proteins. They also comprise the biochemical means by which cells regulate themselves and each other and carry out their various functions inside an organism.

Recessive: A condition that is apparent only when an individual inherits the same allele from both parents; requiring two copies of a certain allele to be expressed. (In his classic breeding experiments in plants, long before the definition of the word *gene* was clear, Gregor Mendel noted that certain traits tended to "recede" in the presence of what he called a "dominant" factor. Hence he coined the term *recessive*.)

Restriction enzyme: A protein, usually derived from a bacterium, which recognizes a specific short sequence of DNA bases and biochemically severs any DNA sequence containing that exact combination. Many geneticists use restriction enzymes as the molecular equivalent of scissors, to cut DNA sequences at will.

RFLP: Restriction-fragment-length polymorphism, a variation in the size of DNA segments cut by restriction enzymes in different individuals within the same population. Molecular geneticists can use many of these variations to specify locations on genetic linkage maps.

Sex chromosomes: The X and Y chromosomes that determine an individual's gender. In humans and most animals (there are some interesting exceptions), two X chromosomes create a female individual and an XY genotype makes a male.

Vector: A DNA molecule originating from a virus, bacterium, or other cell into which genetic material of another origin can be introduced. Molecular biologists use these vectors to transfer foreign DNA segments into the genomes of cells, where they can be studied and preserved.

Virus: The smallest type of organism, nothing more than genetic material wrapped in a protein coat. A virus cannot reproduce independently. It must infect another kind of cell and take over the cell's genetic machinery in order to produce the proteins encoded by its genes and to replicate itself.

BIBLIOGRAPHY AND SOURCES

A comprehensive bibliography relevant to the human genome project might be as thick as this book. For almost any human disease, condition, or even quality, there is likely to be either a substantial literature relating to its heredity or research in progress toward defining its genetic basis. The field of human genetics is burgeoning and changing rapidly. With apologies, I refer any very enthusiastic readers to a scientific literature data base such as MEDLINE.

What follows is a very selective list of references of possible interest to a reader who is not a scientist. Having completed *Mapping Our Genes*, such a reader ought to be able to gain something even from the articles listed below that appear in scientific journals.

I have tried to limit the list to easily accessible journals, and in general to articles that relate to the topics I have chosen to highlight between these covers. For that reason, landmark articles on the subjects of Alzheimer's disease, atherosclerosis, forms of cancer other than colon cancer, and many others are not included.

Scientific Reports, Reviews, and Editorials

BODMER, W. F., BAILEY, C. J., ET AL. "Localization of the Gene for Familial Adenomatous Polyposis on Chromosome 5." *Nature* 328 (August 13, 1987): 614.

BOTSTEIN, DAVID, WHITE, RAYMOND L., ET AL. "Construction of a Genetic Linkage Map in Man Using Restriction Fragment Length Polymorphisms." *American Journal of Human Genetics* 32 (1980): 314.

CANNON-ALBRIGHT, LISA, SKOLNICK, MARK, ET AL. "Common Inheritance of

Susceptibility to Colonic Adenomatous Polyps and Associated Colorectal Cancers." *New England Journal of Medicine* 319:9 (September 1, 1988): 533.

CASKEY, C. T. "Disease Diagnosis by Recombinant DNA Methods." *Science* 236 (June 5, 1987): 1223.

CRAWFORD, M. D'A. "Prenatal Diagnosis of Common Genetic Disorders: Expanding Fast." *British Medical Journal* 297 (August 20–27, 1988): 502.

DONIS-KELLER, HELEN, GREEN, PHILIP, ET AL. "A Genetic Linkage Map of the Human Genome." *Cell* 51 (October 23, 1987): 319.

DULBECCO, RENATO. "A Turning Point in Cancer Research: Sequencing the Human Genome." *Science* 231 (March 7, 1986): 1055(2).

EGELAND, JANICE A., GERHARD, DANIELA S., ET AL. "Bipolar Affective Disorders Linked to DNA Markers on Chromosome 11." *Nature* 325:6107 (February 26, 1987): 783.

GUSELLA, JAMES F., WEXLER, NANCY S., ET AL. "A Polymorphic DNA Marker Genetically Linked to Huntington's Disease." *Nature* 306 (November 17, 1983): 234.

HOFFMAN, ERIC P., FISCHBECK, KENNETH H., ET AL. "Characterization of Dystrophin in Muscle-Biopsy Specimens from Patients with Duchenne's or Becker's Muscular Dystrophy." *New England Journal of Medicine* 318:21 (May 26, 1988): 1363.

JAWORSKI, M. A., SEVERINI, A., ET AL. "Inherited Diseases in North American Mennonites: Focus on Old Colony (Chortitza) Mennonites." *American Journal of Medical Genetics* 32, 2 (February 1989): 158–68.

KEATS, B. J. "Linkage Studies of Friedreich's Ataxia." *American Journal of Human Genetics* 41(4) (October 1987): 627.

KUNKEL, LOUIS M., ET AL. "Analysis of Deletions in DNA from Patients with Becker and Duchenne Muscular Dystrophy." *Nature* 322 (July 3, 1986): 73.

LEDLEY, FRED. "Somatic Gene Therapy for Human Disease: Background and Prospects, Part II." *Journal of Pediatrics* 110, 2 (February 1987): 167.

LEPPERT, M., DOBBS, M., ET AL. "The Gene for Familial Polyposis Coli Maps to the Long Arm of Chromosome 5." *Science* 238 (December 4, 1987): 1411.

MCKUSICK, VICTOR A. "The Morbid Anatomy of the Human Genome: A Review of Gene Mapping in Clinical Medicine." A four-part series in *Medicine*: vol. 65 (1986): 1–33; vol. 66 (1987): 1–63, 237–96; and vol. 67 (1988): 1–19.

MARTIN, JOSEPH B. "Molecular Genetics: Applications to the Clinical Neurosciences." *Science* 238 (November 6, 1987): 765.

ORKIN, STUART H., LITTLE, PETER F. R., ET AL. "Improved Detection of the Sickle Mutation by DNA Analysis: Application to Prenatal Diagnosis." *New England Journal of Medicine,* 307:1 (July 1, 1982): 32.

ROTTER, JEROME I., and RIMOIN, DAVID L. "The Genetics of Diabetes." *Hospital Practice* (May 15, 1987): 79.

ROWLAND, LEWIS P. "Dystrophin: A Triumph of Reverse Genetics and the End of the Beginning" (editorial). *New England Journal of Medicine* 318:21 (May 26, 1988): 1392.

SMITH, LLOYD, and HOOD, LEROY. "Mapping and Sequencing the Human Genome: How to Proceed." *Bio/Technology* 5 (September 1987): 933.

TSUI, LAP-CHEE, BUCHWALD, MANUEL, ET AL. "Cystic Fibrosis Locus Defined by a Genetically Linked Polymorphic DNA Marker." *Science* 230 (November 29, 1985): 1054.

VOGELSTEIN, BERT, FEARON, ERIC R., ET AL. "Allotype of Colorectal Carcinomas." *New England Journal of Medicine* 319:9 (September 1, 1988): 207.

WHITE, RAY, LEPPERT, MARK, ET AL. "Construction of Linkage Maps with DNA Markers for Human Chromosomes." *Nature* 313:101 (January 1985): 10.

WHITE, RAY, and CASKEY, C. THOMAS. "The Human as an Experimental System in Molecular Genetics." *Science* 240 (June 10, 1988): 1483.

WYMAN, A. R., and WHITE, R. "A Highly Polymorphic Locus in Human DNA." *Proceedings of the National Academy of Sciences* 77 (November 1980): 6754–58.

Journalistic or "Popular" Accounts

BARNES, DEBORAH M. "Troubles Encountered in Gene Linkage Land." *Science* 243 (January 20, 1989): 313.

CANTOR, CHARLES R. "Huntington's Disease: Charting the Path to the Gene." *Nature* 308 (March 29, 1984): 404.

KOLATA, GINA. "Two Disease-Causing Genes Found." *Science* 234 (November 7, 1986): 669.

LEWIN, ROGER. "Molecular Biology of Homo Sapiens." *Science* 233 (July 11, 1986): 157.

"Manic-Depression Gene Tied to Chromosome 11." *Science* 235 (March 6, 1987): 1140.

MARX, JEAN L. "Multiplying Genes by Leaps and Bounds." *Science* 240 (June 10, 1988): 1408.

ROBERTS, LESLIE. "Genome Mapping Goal Now in Reach." *Science* 244 (April 28, 1989): 424.

SCHMECK, HAROLD. "Burst of Discoveries Reveals Genetic Basis for Many Diseases." *The New York Times*, March 31, 1987, sec. C, p. 1.

WEINBERG, ROBERT A. "To Sequence or Not to Sequence." *Scientific American* (November 1988): 150.

WHITE, RAY, and LALOUEL, JEAN-MARC. "Chromosome Mapping with DNA Markers." *Scientific American* 258:2 (February 1988): 40.

Monographs and Books

CHERFAS, JEREMY. *Man Made Life*. New York: Pantheon Books, 1983. A very readable history and explanation of biotechnology and genetic engineering.

HOLTZMAN, NEIL A. *Proceed with Caution: Predicting Genetic Risks in the Recombinant DNA Era*. Baltimore, Md.: Johns Hopkins University Press, 1989. A formidable analysis of the practical and ethical implications of genome mapping, preceded by a highly detailed and technical description of the technology and methods.

"Mapping and Sequencing the Human Genome." Washington, D.C.: National Academy Press, 1988. The report of the National Research Council Committee on Mapping and Sequencing the Human Genome (which strongly endorsed the idea). Written for the general public.

"Mapping Our Genes—The Genome Projects; How Big, How Fast?" Congress of the United States, Office of Technology Assessment, April 1988. An OTA feasibility study about the genome project, available from the Superintendent of Documents, U.S. Government Printing Office, Washington, D.C. 20402–99325.

PINES, MAYA. "Mapping the Human Genome." Bethesda, Md.: Howard Hughes Medical Institute, 1987. A monograph on the subject for the general public. Available from HHMI Executive Offices, 6701 Rockledge Drive, Bethesda, Md. 20817.

Interviews

ANTONARAKIS, STYLIANOS. Geneticist, the Johns Hopkins University, Baltimore: brief interview at the Jackson Laboratory's short course on genetics, Bar Harbor, Maine, July 21, 1988.

BENEDICT, WILLIAM. Childrens Hospital of Los Angeles: by telephone, August 23, 1989.

BENNETT, DR. PETER. Director of long-standing study of non–insulin-dependent (adult-onset) diabetes among Pima Indians, Phoenix Indian Medical Center, Arizona: by telephone, August 8, 1988.

BEST, DR. LYLE. Family physician with the Indian Health Service in Rolette, North Dakota: Numerous conversations in Bar Harbor, Maine, during the week of July 18, 1988. I was very disappointed not to be able to include Dr. Best, who is a fascinating man doing very interesting studies of Beckwith-Wiedemann syndrome in an American Indian kindred. The technical problems in the search for a genetic marker using his blood samples, at the time of our last conversation, by telephone on November 17, 1988, made the study's prospects very murky.

BOSWELL, DOROTHYE. Executive director of the National Association for Sickle Cell Disease: by telephone, December 5, 1988.

BOTSTEIN, DAVID. Molecular geneticist, Genentech, Inc., San Francisco: by telephone, December 12, 1988, and follow-up by telephone, August 14, 1989.

BRANDT, JASON. Psychologist, Huntington's disease clinic, Johns Hopkins University School of Medicine, Baltimore: interview in his office, April 7, 1989.

BURT, DR. RANDALL. Gastroenterologist at the University of Utah Health Sciences Center, Salt Lake City: interview in his office, October 10, 1989.

CAHILL, DR. GEORGE. Vice president for scientific training and development, Howard Hughes Medical Institute, Bethesda, Maryland: by telephone, on or about November 1, 1988.

CANNON-ALBRIGHT, LISA. Biostatistician working with Mark Skolnick, University of Utah Health Sciences Center, Salt Lake City: interview in her office, November 7, 1988.

CANTOR, CHARLES. Biophysicist and molecular geneticist, Columbia University department of genetics and development and later director, Human Genome Center, Lawrence Berkeley Laboratory, Berkeley, California: by telephone, August 12, 1988. At laboratory meetings August 21, September 15 and 22, October 3, November 28, December 3, 8, 15, and 22, 1988, and January 24, 1989. Other visits to the lab on April 14 and 15, 1989. Brief interview at the Arden House retreat north of New York City, on September 30, 1988; lengthy interview in his office at the Columbia University Medical Center on June 22, 1989.

CASKEY, DR. C. THOMAS. Molecular geneticist, Baylor College of Medicine, Houston, Texas: at a conference at the National Institutes of Health on November 30, 1988, and again at a conference in Washington, D.C., on April 24, 1989.

CHAMBERLAIN, SUSAN. Molecular geneticist, St. Mary's Hospital Medical School, London: interview during the American Society of Human Genetics meeting in New Orleans, October 13, 1988.

CHENG, JAN-FANG. Molecular geneticist, postdoctoral fellow in Cassandra Smith's lab at Columbia University: occasional conversations and interviews, most notably December 22, 1988.

COLLINS, DR. FRANCIS. Physician and molecular geneticist, University of Michigan, Ann Arbor: by telephone October 6, 1988, and June 5, 1989, at the American Society of Human Genetics meeting in New Orleans, October 13, 1988, and at a conference at the National Institutes of Health on November 30, 1988, and during and after a Washington, D.C., press conference announcing the discovery of the cystic fibrosis gene, August 24, 1989.

COMINGS, DR. DAVID. Physician and geneticist, City of Hope Medical Center, Duarte, California: breakfast interview during the American Society of Human Genetics annual meeting in New Orleans, October 15, 1988.

COMO, PETER. Neuropsychologist, University of Rochester Medical Center: in Holland, Michigan, on October 20, 1988. Also throughout the field trip to Holland, October 20 through 23, 1988. Checking interview, by telephone, February 3, 1989.

COOK-DEEGAN, DR. ROBERT. Acting executive director, Biomedical Ethics Board, advisory committee to the U.S. Congress, by telephone, January 11, 1988, and at a conference in Washington, D.C., on April 25, 1989.

DAUSSET, JEAN. Immunologist, director of the Centre d'Etude du Polymorphisme Humain, Paris: by telephone to his summer house in Spain, September 18, 1989.

DOGGETT, NORMAN. Molecular geneticist, postdoctoral fellow in Cassandra Smith's Columbia University laboratory: brief interviews, December 15, 1988, and February 24, 1989.

DONIS-KELLER, HELEN. Geneticist, formerly of Collaborative Research, Inc., later of Washington University, St. Louis: by telephone, May 4, 1989.

EGELAND, JANICE. Professor of Psychiatry, University of Miami, School of Medicine, Miami, Florida: by telephone, December 1981, and at a grand rounds at the New York State Psychiatric Institute, December 4, 1981; by telephone, July 2, 1987; again by telephone, October 13, 1987, to request permission to write a book about her research, at which time she invited me to call back the following spring but urged me meanwhile to look into the other important genetic research going on in other ethnic isolates such as the Cajuns and the Navajo. She suggested that this research had the potential of explaining the causes of and "wiping out" many common diseases, much as the germ theory was a breakthrough in the understanding of infectious disease. Interviews by telephone, April 22 and July 17, 1988, and August 14, 1989, to check the historical accuracy of information in the chapter dealing with her research. Also I heard her address accepting the Rema LaPouse Award at the meeting of the American Public Health Association in Boston on November 15, 1988, and we had a brief conversation afterward. Checking interview, by telephone, September 11, 1989.

ELSTON, ROBERT. Chairman of biometry, Louisiana State University Medical Center, New Orleans: in his New Orleans office, October 19, 1988.

ENDICOTT, JEAN. Psychologist, New York State Psychiatric Institute: interview in her Manhattan office, February 24, 1989.

ERNEST, LAURIE. Nurse, University of Rochester Medical Center: interview at lunch, October 20, 1988, and throughout field trip to Holland, Michigan. Also by telephone, February 5, 1989.

FINE, BETH. Genetic counselor and former president of the National Society of Genetic Counselors, Illinois Masonic Medical Center, Chicago: by telephone, September 2, 1988, and during a coffee break at the American Society of Human Genetics meeting in New Orleans, October 12, 1988.

FISCHER, STUART. Biophysicist, Lawrence Berkeley Laboratory Human Genome Project, Berkeley, California: interview in Cantor-Smith laboratory at Columbia University, January 18, 1989.

FREDRICKSON, DR. DONALD. Former president, Howard Hughes Medical Institute, Bethesda, Maryland: by telephone, April 28, 1989.

FRIEDMAN, ORRIE. Chairman and chief executive officer, Collaborative Research, Inc., Bedford, Massachusetts: by telephone, May 4, 1989.

GERHARD, DANIELA. Molecular geneticist, Washington University Medical Center in St. Louis: interview over lunch during the American Society of Human Genetics annual meeting in New Orleans, October 12, 1988. Follow-up conversation in Manhattan, March 6, 1989.

GERSHON, ELLIOTT. Chief of clinical neurogenetics, National Institute of Mental Health: interview in his office, November 29, 1988. I was very late and he was very gracious. Interview was completed by telephone, December 7, 1988.

GESTELAND, RAY. Geneticist, University of Utah, Salt Lake City: interview in his office, October 10, 1988.

GILBERT, WALTER. Molecular biologist, Harvard University, Cambridge, Massachusetts: by telephone, October 1, 1987, about his efforts to set up a company with the goal of sequencing the human genome.

GOLDMAN, DAVID. Chief of the genetics study section, National Institute of Alcohol Abuse and Alcoholism: about proposed genetic studies of alcoholism in native American Indian tribes, by telephone, August 12 and 18, 1988.

GOLLER, EVELYNE PITRE. Betty LeBlanc's cousin: interview in her house in Breaux Bridge, Louisiana, October 15, 1988. Correspondence dated January 28, and February 18, 1988.

GUSELLA, JAMES. Molecular geneticist, Massachusetts General Hospital, Boston: interview in his Boston office, November 15, 1988, and a response to further questions sent by mail, in a letter dated February 24, 1989.

HOSTETLER, JOHN. Anthropologist/sociologist, director of the Center for the Study of Anabaptist and Pietist Groups, Elizabethtown College in Pennsylvania: interview in his office at Elizabethtown College, February 20, 1989.

HOSTETTER, ABRAM. Psychiatrist in private practice in Hershey, Pennsylvania: interview in his office and at a Hershey restaurant over lunch on August 26, 1988.

HOUSMAN, DAVID. Molecular geneticist, Massachusetts Institute of Technology, Cambridge: interview in his office, August 8, 1988.

JANZEN, DAVID. Unemployed Mennonite man diagnosed with Tourette's syndrome: interview in the kitchen of his mobile home in La Crete, Alberta, Canada, on September 10, 1988.

JAWORSKI, DR. MARTINE. Formerly pediatrician and immunologist, University of Alberta, Canada, and now deputy medical director, the Canadian Red Cross Society: by telephone, July 13, 1987; by telephone, August 4, September 19, October 8 and 19, 1987; January 6 and 11, November 22, and December 6 and 13, 1988. Interviews and visits to Edmonton laboratory, September 7 and 14, 1988, in her Edmonton home, September 15, 1988; by telephone, January 13, March 15, March 20, April 15 and 17, and August 15, 1989.

KAN, DR. YUET WAI. Hematologist and geneticist, University of California at San Francisco: by telephone, December 19, 1988.

KEATS, BRONYA. Biometrician (genetic statistician), Louisiana State University Medical Center, New Orleans: by telephone, January 6 and 11, 1988; at her laboratory and at a restaurant in New Orleans during lunch on October 17, 1988; brief follow-ups by telephone, March 15 and May 25, 1989.

KIDD, KENNETH. Molecular geneticist, Yale University School of Medicine, New Haven, Connecticut: interview in his office, January 23, 1989; follow-up interview by telephone, January 26, 1989.

KLIMUSKA, ED. Reporter who has the "Amish beat" for the *Lancaster New Era:* conversation at the newspaper office, in Lancaster, Pennsylvania, August 26, 1988.

KUNKEL, LOUIS. Molecular geneticist, Boston Children's Medical Center: interview August 9, 1988.

KURLAN, DR. ROGER. Neurologist, University of Rochester: by telephone, July 1, 1987, and August 24, 1988; checking interviews by telephone, January 27 and August 14, 1989. Also in Holland, Michigan, at lunch on October 20, 1988, and at breakfast on October 23, 1988, and throughout the field trip between October 20 and 23, 1988.

In the last telephone interview, Dr. Kurlan urged me in the strongest terms not to use the real names of the Van Dyke and Holbein families, even though I had obtained signed releases from the members of the families most closely involved in his studies. If I used real names, he argued, other members in the family would by definition be identified as at risk of Tourette's syndrome without their permission and perhaps against their will.

This is a major problem in reporting about genetic diseases, and one my profession must begin to confront. I decided to do as he asked. (Because the LeBlanc family had already appeared prominently in Texas and Louisiana newspaper articles about Friedreich's ataxia, and because Betty LeBlanc actively wishes to confront the issue of a misguided stigma about the disease, I did not feel constrained in the same way in their case.)

LALOUEL, DR. JEAN-MARC. Physician and genetics statistician, University of Utah Health Sciences Center, Salt Lake City: interview in his office, October 10, 1988.

LEBLANC, BETTY. Texas housewife and mother of three children with Friedreich's ataxia: by telephone, January 6, 12, and 26, February 10, August 23 and December 19, 1988, and January 10, January 27, and April 27, 1989. Interviews at her home, October 8 through 11, 1988. (Her husband Derald was also involved in many of these conversations.) Correspondence dated January 8 and September 1988, January 7, February 23, and July 28, 1989. Mrs. LeBlanc was tireless and enthusiastic in providing me with letters and other documents relating to her family's history and her son's medical treatment.

LEBLANC, DARREN. Biochemistry student affected with Friedreich's ataxia: interview in his apartment at Lamar University, Beaumont, Texas, October 10, 1988. Correspondence dated August 7, 1989.

LEBLANC, DAVID. Darren's older brother: in his kitchen in Orange, Texas, October 8, 1988.

LEDER, DR. PHILIP. Physician and geneticist, Harvard University, Cambridge, Massachusetts: by telephone, November 4, 1987.

LEFEVER, DON. Public relations officer, Church of Jesus Christ of Latter-Day Saints, Salt Lake City: by telephone, July 6, 1987.

LEPPERT, MARK. Geneticist, University of Utah Health Sciences Center, Salt Lake City: interview in his office, October 10, 1988.

LOCKE, BENJAMIN. Chief of epidemiology and psychopathology research, National Institute of Mental Health: by telephone, February 14, 1989.

LOWRANCE, WILLIAM. Biomedical ethicist, Rockefeller University, New York City: interview, August 10, 1988.

MCKUSICK, DR. VICTOR. Cardiologist and medical geneticist, the Johns Hopkins University Medical Center, Baltimore, Maryland: by telephone, August 3, 1987; at a National Research Conference seminar on the genome project held in Washington, D.C., February 11, 1988; by telephone, July 13, 1988; at the Jackson Laboratory's short course on genetics in Bar Harbor, Maine, July 20, 1988; and a long interview in his Baltimore office, November 30, 1988.

MEYERS, DEBORAH. Genetic biostatistician, the Johns Hopkins University Medical Center, Baltimore, Maryland: by telephone, August 12 and August 18, 1988.

MICHELS, BILL. Technician in Cassandra Smith's Columbia University laboratory: lengthy interview about basic laboratory techniques, September 26, 1988. Other conversations in the laboratory December 14, 20, and 22, 1988, and March 2, 1989.

MOLIVAR, RICHARD. Principal investigator, National Genetic Mutant Cell Repository, Camden, New Jersey: by telephone, February 17, 1989.

MOTULSKY, ARNO. Medical geneticist, University of Washington at Seattle: during a coffee break at the American Society of Human Genetics meeting in New Orleans, October 13, 1988.

MURRAY, DR. ROBERT F., JR. Internist and medical geneticist, Howard University School of Medicine, Washington, D.C.: during a coffee break at the Washington conference, April 25, 1989.

MURRAY, THOMAS. Biomedical ethicist, Case Western Reserve University, Cleveland, Ohio: at dinner in Washington, during a conference about the genome project, April 24, 1989.

NAKAMURA, DR. YUSUKE. Surgeon and molecular geneticist, University of Utah Health Sciences Center, Salt Lake City: interviews in his laboratory on November 8 and 10, 1988, and at dinner on November 8. In 1989, Nakamura returned to Japan.

NASH, JEAN. Director, Resource for Genetic and Epidemiologic Research, University of Utah Health Sciences Center, Salt Lake City: interview in her office, October 10, 1988.

NEE, LINDA. Social studies analyst at the National Institute of Neurological Diseases and Stroke, about her studies of the genetics of Alzheimer's disease: by telephone, October 19, 1987.

NEEL, DR. JAMES. Physician and human geneticist, University of Michigan Medical Center, Ann Arbor: by telephone, December 5, 1988.

NEPOM, GERALD. Immunogeneticist, University of Washington School of Medicine, Seattle: by telephone, March 20, 1989.

NEUFELD, WILLIAM. Farmer and amateur local historian of La Crete, Alberta, Canada: interview in his kitchen, September 11, 1988. Mr. Neufeld kindly showed me numerous historic documents he had obtained in preparing a history of the community.

PAPOLOS, DEMITRI. Psychiatrist, Albert Einstein Medical Center, New York: by telephone, August 18, 1988, and interview in his office, October 4, 1988.

PAULS, DAVID. Human geneticist, Child Study Center, Yale University School of Medicine, New Haven, Connecticut: by telephone, July 2, 1987. Interview September 28, 1988, in his office and at American Society of Human Genetics meeting in New Orleans, October 13, 1988.

PELIAS, MARY KAY. Geneticist, Louisiana State University Medical Center, New Orleans: by telephone, January 6, 1988, and at lunch in a New Orleans restaurant on October 17, 1988.

RINGQUIST, STEVE. Molecular geneticist, postdoctoral fellow in Cassandra Smith's Columbia University laboratory: interview over lunch in his apartment, December 15, 1988.

RODGERS, DR. GRIFFIN. Senior staff fellow of the National Institute of Diabetes and Digestive and Kidney Diseases, studying sickle-cell anemia: interview in the library at the NIH Clinical Center, November 29, 1988.

ROWLEY, DR. PETER. Medical geneticist, University of Rochester School of Medicine, Rochester, New York: by telephone, October 25, 1988.

SAITO, DR. AKIHIKO. Physician and postdoctoral fellow in molecular genetics at the Cantor-Smith lab at Columbia University until mid-1989: Observation and interviews in the laboratory on December 3, 6, 8, 9, 14, 15, and 20, 1988, and January 18 and 24, 1989. Interview at a restaurant in New Orleans, during the American Society of Human Genetics meeting, October 14, 1988. Casual interview at Aki's farewell party, March 2, 1989.

SCOTT, STEVE. Researcher and writer, People's Place, Intercourse, Pennsylvania (a research and public information center and tourist attraction devoted to Amish culture): interview at People's Place, February 21, 1989.

SEVERINI, ALBERTO. Biochemist, University of Alberta: interviews in his laboratory September 7, 8, and 14, 1988; by telephone November 23, 1988, and March 17, 1989.

SHARP, PHILLIP. Director, Center for Cancer Research, Massachusetts Institute of Technology, Cambridge: on the front porch of a guesthouse in Bar Harbor, Maine, and later over dinner at a restaurant on July 18, 1988, we had a lengthy and (to me) most heartening conversation about the form of this book and people who might be willing to help.

SIMMONS, JOHN. Biologist, Utah State University, Logan, Utah: about the late Eldon Gardner, by telephone, April 27 and 28, 1989.

SKOLNICK, MARK. Population geneticist, University of Utah Health Sciences Center: in his Salt Lake City office on November 7, 1988. He refused to be taped or officially interviewed in person but talked at great length anyway and allowed me to take notes for background information. Skolnick agreed instead to respond to questions sent to him by mail. He responded to one set of questions by letter on February 25, 1989; I received no response to a set of follow-up questions, but he did review a draft of the chapter for accuracy and suggested some additions, which are included.

SLATER, DR. JONATHAN. Pediatrician, University of Alberta Medical Center, Canada: in the medical center cafeteria, September 14, 1988.

SMITH, CASSANDRA. Molecular biologist, Columbia University: two interviews, at a restaurant in Fort Lee, New Jersey, on August 29, 1988, and April 10, 1989, as well as numerous brief conversations during visits to her laboratory as listed above at Charles Cantor.

THURMON, DR. THEODORE F. Physician and geneticist, Louisiana State University Medical Center, Shreveport: by telephone, October 19, 1987. Dr. Thurmon is studying hereditary diseases in certain very isolated triracial populations in western Louisiana.

VINCENT, CLYDE. Betty LeBlanc's cousin, an amateur genealogist of his own family: in the dining area of his house in Beaumont, Texas, October 9, 1988. Clyde also drove me to a nearby park, where he proudly showed me the main object of his most recent efforts: the relocation and restoration of an authentic Cajun farmhouse as a historic site.

VINCENT, PEARLY. Betty LeBlanc's mother: interview in her house in Morse, Louisiana, October 16, 1988. Betty's father was there too, but didn't say very much.

VOGELSTEIN, BERT. Cancer geneticist, the Johns Hopkins University School of Medicine, Baltimore, Maryland: by telephone, May 2, 3, and 4, 1989.

WATSON, JAMES. Geneticist, director of Cold Spring Harbor Laboratory, New York, and also of the National Institutes of Health's human genome project: interview in his Cold Spring Harbor office, June 15, 1989.

WEAVER, ZOE. Microbiology student, University of Michigan, Ann Arbor: by telephone, September 20, 1989.

WEINBERG, ROBERT. Geneticist, Whitehead Institute for Biomedical Research,

Cambridge, Massachusetts: by telephone, October 1, 1987, about proposals
to sequence the human genome.

WEXLER, NANCY. Psychologist, Columbia University and president of the Hereditary Disease Foundation: by telephone, August 1, 1988, and numerous
cordial conversations over coffee at genetics meetings. I drew heavily on her
address to the AMA Alliance for Aging Research conference in Washington,
D.C.

     The lack of a lengthy personal interview with Wexler may seem re-
markable. She is a pivotal figure in the dialogue about these issues and in
facilitating genetic research. But our schedules gridlocked repeatedly, and
by the time I got around to setting a date, I realized that I had already gained
so much from hearing her speak, reading her eloquent words, and hearing
others talk about her influence that it seemed I could gain little more by
wedging myself into her busy schedule for a private interview.

WHITE, RAYMOND. Geneticist, University of Utah School of Medicine, Salt Lake
City: by telephone, July 15, 1987; on the grounds of Jackson Laboratory, in
Bar Harbor, Maine, July 20, 1988. Interviews in his laboratory on Novem-
ber 7, 10, and 11, 1988, and at lunch in Salt Lake City on November 9, 1988.
Sat in on a lecture to an undergraduate genetics course, University of Utah,
November 8, 1988. Follow-up interview by telephone, May 25, 1989.

WHITTEN, DR. CHARLES. Director of the Comprehensive Sickle-Cell Center at
Wayne State University, Detroit, Michigan, and president, National Association for Sickle Cell Disease: by telephone, December 5, 1988.

WILENSKY, DR. MICHAEL. Neurologist, Kenner, Louisiana: by telephone, October 19, 1988.

WILLIAMSON, ROBERT. Molecular geneticist, St. Mary's Hospital Medical School,
London: Conversations at the American Society of Human Genetics meet-
ing, October 13, 1988, and at a meeting on gene mapping sponsored by the
National Heart, Lung, and Blood Institute in Bethesda, Maryland, on
November 30, 1988.

WOOD, DONALD. Biochemist, research director of the Muscular Dystrophy Association: interview in his Manhattan office, September 29, 1988.

ZINDER, NORTON. Geneticist, Rockefeller University, New York City, and chairman of the advisory committee to the U.S. Human Genome Initiative: by
telephone, December 6, 1989.

Anonymous Interviews

Perhaps no writer is very happy using pseudonyms or gaining information off
the record. Many journalists will never do so. I had several reasons to overrule
my own discomfort about this.

     In some cases, I did not want to violate the privacy of those members of
hereditary-disease kindreds who had not talked to me and did not know I or
this book exist. In the peculiar case of the Amish, their own ethic prohibits
them from gaining publicity of any kind; as Janice Egeland found long before
me, no one will cooperate except on a confidential basis.

     Sometimes off-the-record interviews also helped me to gain insights that were useful in depicting people or events. I never made a factual statement in this book based on an off-the-record assertion. I have merely used the statements to balance or substantiate impressions gained from others on the record. The nurses in La Crete talked on the record, but I felt they might be uncomfortable if I used their names, and I saw no good reason to do so.

Four nurses working at the public health unit in La Crete, Alberta: interviews at the unit on September 9, 10, 11, and 12, 1988, and by telephone on March 21, 1989. I came to have the highest regard for all of them, and became very fond of one in particular, who did me the favor of giving me dinner at her farmhouse while her husband and sons were using the combine late into the night to harvest the canola before it rained.

"Frances," a lawyer and now housewife, member of a family affected with Huntington's disease: in the living room of her suburban home, April 20, 1989.

"Holbein" family: affected by Tourette's syndrome, in Holland, Michigan, October 20–23, 1988.

"Jakey's" mother, a Mennonite farmwife: interview at her kitchen table in her farmhouse outside La Crete, on September 12, 1988. I interviewed "Jakey's" aunt, the mother of his cousin who also has diabetes, in her La Crete home, on the same day.

"Mary Stoltzfus," single woman and shopworker in Lancaster County, under treatment for bipolar disorder: interview in my car as she ate her sandwich during her lunch break on February 22, 1989. I am very grateful for her candor.

A member of the NIH grants review committee reviewing Janice Egeland's grant proposals: February 17, 1989.

Two scientists who have worked briefly in the Cantor-Smith laboratory but have since left also granted me interviews on a confidential basis. I also followed "Nikolai," a pseudonymous postdoc in the lab who was working on PCR and on linking clones, but decided not to write much about him or (as he requested) to use his real name, which might jeopardize his career.

Unnamed Amish grandfather and grandmother: well acquainted with Janice Egeland and her research, graciously allowed me to stay in their farmhouse outside Intercourse and have lengthy conversations about a number of topics, February 20 through 23, 1989. During the course of the visit I also had supper with a daughter's family and visited some of their Amish friends.

Unnamed single Amish woman, involved with Janice Egeland in her early research: at her kitchen table, February 22, 1989.

"Van Dyke" family: parents and other relatives of Mrs. "Van Dyke," also affected by Tourette's syndrome, at the "Van Dyke" home and at a nursing home in and near Holland, Michigan, October 20–23, 1989.

Conferences and Seminars

Annual Meeting, the American Society of Human Genetics, New Orleans, October 12–15, 1988.

"Genetic Basis of Human Disease: Molecular Mechanisms and Strategies for Therapy": a conference sponsored by the National Heart, Lung, and Blood Institute, Bethesda, Maryland, November 30 and December 1, 1988.

National Academy of Sciences, public briefing on the human genome project, Washington, D.C., February 11, 1988.

Short Course in Medical and Experimental Mammalian Genetics, Jackson Laboratory, Bar Harbor, Maine, July 18–23, 1988.

"Unlocking Potential: The Promise of the Human Genome Initiative": a conference sponsored by the Alliance for Aging Research and the American Medical Association, and funded by the E. I. du Pont de Nemours & Co., Inc., Washington, D.C., April 24 and 25, 1989.

INDEX

The letter *f* following a page number means "figure," or illustration.